WILDLIFE QUIZ
& AMAZING FACTS BOOK

FUN AND FASCINATING FACTS FOR ALL

Don Conroy / Chris Wilson

Editor: Alan McGuire

NATURAL RAPTURE

This First Edition Published by: Natural Rapture Ltd.
1999
A Natural Rapture Ltd. Publication,
The Studio,
Ballinamorragh, Curracloe, Co. Wexford, Ireland.

All rights reserved. No part of this publication may be reproduced, stored in a retrieval system or transmitted, in any form or by any means, electronic, mechanical, photocopying, recording, or otherwise, without prior written consent of the copyright owners.

© Conroy / Wilson

Written by Don Conroy and Chris Wilson.

ISBN 0 9537208 0 2

Design and Origination by Phototype-Set Ltd.,
2 Lee Road, Dublin Industrial Estate, Glasnevin, Dublin 11.

Printed by ColourBooks Ltd.
105 Baldoyle Industrial Estate, Baldoyle, Dublin 13.

Cover Design: Natural Rapture Ltd.

Photography and Illustrations: Don Conroy / Chris Wilson

Introduction

Over the years we have both had the pleasure of meeting and holding workshops with many young people in schools, libraries and community halls throughout the country. The enthusiasm of these young people, and their fascination with wildlife is most encouraging and offers great hope for the future. In our travels we are often asked some amusing, and often challenging, questions about wildlife. How far can a swallow fly? How many eyes has a spider? What is the largest mammal in the world? This led us to the idea of putting this book together and cramming it with quizzes and amazing facts about our natural world. It turned out to be a bigger task than we had imagined, but as we both have a passion for all aspects of natural history, we took on the task and thoroughly enjoyed the research. Here we must thank all our friends and colleagues for their help and encouragement. Our biggest challenge was to produce a book that would appeal to all ages, that would intrigue, inform and entertain. Here is the result. We hope that you'll enjoy it and, if you cannot answer some of the questions, we hope you'll enjoy checking out the answers.

Don Conroy
Christopher J. Wilson

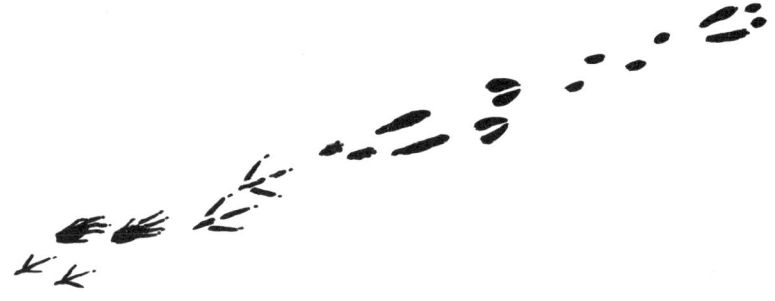

Wildlife Quiz

1. What is a capercaillie?
2. Where would you find the Caledonian forest?
3. What is a pachyderm?
4. Where would you find the wintergreen plant growing?
5. Where would you see wildcats?
6. What is a pintail?
7. What is a bogbean?
8. What is a sundew?
9. What is Ireland's largest duck?
10. What are young spiders called?
11. What are the seeds of an ash tree called?
12. Are mute swans mute?
13. What is a wheatear?
14. In Bram Stoker's novel *Dracula*, what creatures are called 'the children of the night'?
15. Where did the Bewick's swan get its name?

ANSWERS

1. A turkey-sized game bird found in the forests of Scotland and Europe.
2. This last remaining native conifer forest of Britain is found in Scotland.
3. A thick-skinned mammal such as the elephant, hippopotamus and rhinoceros.
4. This single-flowered plant is found amongst the thick carpet of pine needles in conifer forests.
5. Larger than the domestic cat, these wild animals live in the Scottish highlands and the mountainous regions of Europe.
6. A beautiful duck that, as its name suggests, has pointed tail feathers.
7. An attractive wetland plant.
8. An insectivorous plant that traps insects on its sticky hairs. Found in bogs and heaths.
9. The shelduck.
10. Spiderlings.
11. Keys.
12. No, they make a variety of grunting and hissing noises (young are more vocal than adults).
13. A migrant bird belonging to the thrush family.
14. Wolves.
15. They were named after the wood engraver Thomas Bewick who died in 1828. He was famous for his publication of the *History of British Birds*.

Wildlife Quiz

1. Name the three types of swan found in Ireland in winter?
2. Where did our common pheasant originate?
3. Montagu's, hen and marsh are members of which bird family?
4. What is the smallest raptor seen in Ireland and Britain?
5. What is the technical term for a bill?
6. What is the old English name for the mistle thrush?
7. What is the old English name for a grey heron?
8. What is the name given to a nesting colony of grey herons?
9. What is a treecreeper?
10. What is water avens?
11. Why is the drinker moth so called?
12. What plant is the source of the drug digitalin which is used to treat heart disease?
13. What is the largest water beetle in Britain?
14. House and tree indicate which bird family?
15. What is a hobby?

ANSWERS

1. Mute, whooper and Bewick's.

2. From South Russia (in 1589) and from China (in late 18th century), introduced to Ireland and Britain for hunting purposes.

3. The harrier family (birds of prey).

4. The merlin.

5. Rostrum.

6. A throstle, also commonly called a storm cock.

7. A handsaw.

8. A heronry.

9. A small woodland bird with a long curved bill which it uses to probe the bark of trees for insects.

10. A plant that grows near water and has nodding flowers.

11. Because it likes to live in damp habitats.

12. The foxglove, found in woodlands and hedgerows.

13. The great silver beetle reaching almost 50 mm. (2 inches) in size.

14. They are members of the sparrow family.

15. A small falcon, rare in Ireland, that breeds in Britain and Europe. They are particularly adept at feeding on swallows and martin species.

Wildlife Quiz

1. Name the four species of snipe recorded in the British Isles?
2. What is an algal bloom?
3. What is a whip-poor-will?
4. What is a canvasback?
5. What is a sawbill?
6. What does the word 'saxicolous' mean?
7. What does the general term 'tubenose' refer to?
8. What does the expression 'as dead as a dodo' mean?
9. When and where was the last great auk (now extinct) seen in Ireland?
10. When did the great auk become extinct?
11. What does 'diurnal' mean?
12. What does 'nidicolous' mean?
13. What percentage of the earth's surface is covered by water?
14. What does 'nocturnal' mean?
15. What does 'nidifugous' mean?

ANSWERS

1. Common snipe, great snipe, jack snipe and woodcock.
2. A sudden growth of algae in an aquatic ecosystem.
3. A member of the nightjar family found in North America and Mexico. This bird gets its name from its call.
4. An American duck with reddish head, whitish body and black chest. Similar in appearance to the European pochard.
5. A general term for the merganser duck species.
6. It means growing on stones, rocks, walls etc.
7. The group of birds, including albatrosses, fulmars, shearwaters, storm petrels and diving petrels. They all have a particularly well developed olfactory part of the brain and rely heavily on smell for finding food, mates, breeding areas etc.
8. Completely obsolete or defunct. Comes from the fact that the dodo, a large bird formerly found in Mauritius, is extinct.
9. The last Irish great auk record was of a specimen captured alive on the Waterford coast in May 1834.
10. In about 1844.
11. The animal is active during the day time.
12. A term applied to a young bird that stays in the nest until able to fly. When hatched the young are usually naked and unable to move around much.
13. 70%.
14. During the night time.
15. Applies to young birds that on hatching leave the nest almost immediately. They are usually covered in down and able to move about.

Wildlife Quiz

1. What is cuckoo spit?
2. What are nature's common warning colours?
3. How does a grasshopper produce its song?
4. Crested dog's-tail, common fox-tail, false oat and are all species of what?
5. What is an oblong woodsia?
6. What is a clone?
7. What is camouflage on an animal?
8. What is an ecosystem?
9. What is meant by 'fauna'?
10. What is a habitat?
11. What is meant by 'exotic' in the natural world?
12. What is a herbivore?
13. How many legs has an insect?
14. What is an invertebrate?
15. What are marginal plants?

ANSWERS

1. It is the froth surrounding an insect called a froghopper in its nymph stage.
2. Yellow and black.
3. The song is produced by the grasshopper rubbing its back legs (which has tiny pegs) against its wings.
4. They are all species of grass.
5. A fern.
6. An asexually produced offspring of a common ancestor.
7. The natural colouring which enables it to blend into its surroundings.
8. Describes a discrete unit that consists of the living and non-living parts, interacting to form a stable system. It can be used for small as well as huge areas. e.g. A pond or the world.
9. The animal life.
10. The home of a species, organism or community.
11. A species of plant or animal introduced into an area where it would not occur naturally.
12. A plant-eater.
13. Six.
14. An animal without a backbone.
15. Plants that grow at the water's edge.

Wildlife Quiz

1 What is a patoo?
2 What is a capybara?
3 What giant water snake can be found in the Amazon?
4 What is a tragopan?
5 What is a langur?
6 What is the Shell Shiva?
7 What is lava?
8 Where is the Rift Valley?
9 What are stalactites and stalagmites?
10 What is the difference between stalactites and stalagmites?
11 What is a slow worm?
12 Name the three species of snake found in Britain.
13 Name the only species of frog in Ireland.
14 Name the frog species found in Britain.
15 The natterjack toad runs instead of hopping. True or false?

ANSWERS

1   A South American nightjar (bird).
2   A giant rodent and relative of the guinea pig.
3   The green anaconda.
4   A spectacular member of the pheasant family, the size of a turkey, found along the lower regions of the Himalayas.
5   A species of monkey common in India.
6   A collection of fossil ammonites considered by Hindus as creations by the God Shiva.
7   Boiling liquid rock.
8   East Africa.
9   Mineral deposits in caves.
10 Stalactites grow down from the roof of a cave or overhang and stalagmites grow up from the floor of the cave.
11 A legless lizard.
12 The smooth snake, the grass snake and the adder (sometimes known as the common viper).
13 The common frog.
14 The common frog, the edible frog and the marsh frog.
15 True.

Wildlife Quiz

1. What is a nuthatch?
2. Where would you find the harpy eagle?
3. What is the largest penguin in the world?
4. What do polar bears feed on?
5. What is a bobcat?
6. What is a lynx?
7. What is a wolverine?
8. Where would you find the Macaroni penguin?
9. Is the koala a member of the bear family?
10. What do koalas mainly feed on?
11. What is a marsupial?
12. What is a rosella?
13. Where would you find a lyrebird?
14. Where would you find the monkey-eating eagle?
15. What is the most common penguin in the world?

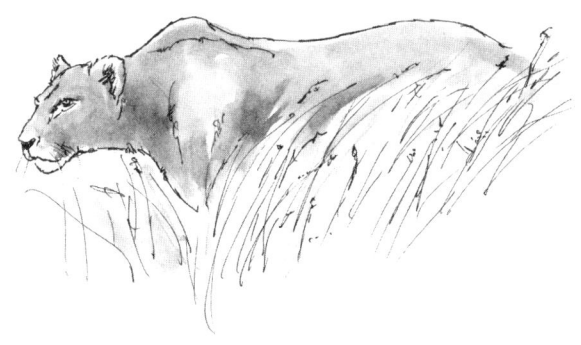

ANSWERS

1 A small bird.
2 South America.
3 The Emperor penguin. They stand waist-high to a man.
4 Mainly seals, small mammals (such as lemmings and foxes), and fish.
5 A large wild cat the size of a collie dog, found in North America.
6 A close relative of the bobcat found in Canada and in the forests of the European continent.
7 The largest member of the weasel family.
8 On Marion Island in the Indian Ocean.
9 No.
10 Eucalyptus leaves.
11 An animal that carries its young in its pouch. Young are born immature and finish development in the mother's pouch.
12 A member of the parrot family.
13 In Australia.
14 The Philippines. It is one of the rarest birds in the world with less than 100 still surviving.
15 The Adélie penguin. They are found only on the Antarctic continent and a few islands close by.

Amazing Facts

Bat Take-Away
Ten million Mexican free-tailed bats live in Bracken Cave. Collectively they remove 100 tons of insects from the sky every night.

Animals On Trial
In less enlightened times animals who broke the laws of man had to suffer the consequences. For instance, in the Middle Ages in France, pigs were allowed to roam freely through village streets, a freedom they sometimes abused. One such pig was hanged in 1394, in Normandy, for eating a child. A similar crime was committed by a sow and her six piglets in 1547. The sow was hanged but the piglets were spared because of their age and the bad example set by their mother.

Inventive Sea Otter
The sea otter uses a 'favourite' stone to crack open shellfish. When the stone is not in use the otter carries it around under its armpit ready for the next occasion.

On The Scent
The bloodhound's sense of smell is one million times more acute than a human's.

Fang Facts
A male baboon's huge fangs are longer than those of a male lion.

Inside Information
Each human body cell contains 1 metre (3 feet 3 inches) of DNA made up of 5 billion base pairs.

Atom Sized
If 100 million atoms are laid end to end, they would equal approximately 2.54 cm. (1 inch).

Mammal Quiz

1. When did the first true mammals appear on earth?
2. Which animal is the fussiest eater?
3. Human vision is better than a cat's. True or false?
4. All mammals give birth to live young. True or false?
5. The weasel family includes stoats, skunks, badgers and otters. How many species are there world-wide?
6. Which is the largest land carnivore?
7. The level of methane gas is increasing in the atmosphere eight times faster than carbon dioxide. Which animal is the main culprit for this?
8. When did the first primates appear?
9. Dogs, both wild and domesticated, have a highly developed sense of smell. How much more efficient is it than a human's?
10. What marine mammal is believed to have inspired the myth of the mermaid?
11. The Indian lion is an endangered species living in the Gir Forest National Park. Approximately how many individuals remain alive?
12. Grey whales spend half of every year travelling from summer feeding grounds in the arctic to breeding sites off California. What distance do they travel?
13. What wild animal likes to use a makeshift umbrella to keep itself dry?
14. Weasels are common in Ireland. True or false?
15. When did the grey wolf become extinct in Ireland?

ANSWERS

1. About 220 million years ago during the Triassic period.
2. The koala. Of the 500 kinds of eucalyptus species of tree in Australia, it only eats the leaves of six. It consumes about 0.5 kg. of leaves a day, yet sifts through over 8 kg.
3. False. All 37 species of cat can see up to six times better in the dark than humans.
4. False. The duck-billed platypus and two species of spiny ant-eater lay eggs.
5. Nearly 70 species. However, you will not find any in Australasia.
6. The polar bear. One animal was recorded as weighing 1,000 kg. (2,200 lbs.) and was 3.4 metres (11 feet) long.
7. The annual methane emissions of domestic cows exceed 62,000 million kgs. (136,000 million lbs.) of gas.
8. 65 million years ago during the Cretaceous period.
9. In tests dogs have been found to be between a thousand and a million times more sensitive to smell than we are.
10. The dugong, a sea cow. Sailors, observing female dugongs nurturing their young, thought they resembled humans with tails.
11. 200.
12. The round trip, which takes three months each way, is about 20,000 kms. (12,500 miles). That is about the average distance a family car covers each year!
13. The orang-utan. It tears off large leaves to shelter beneath during rainforest downpours.
14. False. There are no weasels in Ireland. The Irish stoat is often mistakenly called a weasel.
15. The late 18th century. It is popularly believed that the last wolf present in Ireland was at Ballydarton, Co. Carlow in 1786.

Wildlife Quiz

1. Which big cat is found in the South American jungle. Is it the leopard, the tiger or the jaguar?
2. Which animal is called 'the old man of the woods'?
3. What is a sloth?
4. Which armadillo can roll itself into a ball shape?
5. What is a rhea?
6. What is a takahe?
7. What is a caracara?
8. How many species of penguin are there?
9. What is the biggest land mammal?
10. What is the biggest animal in America?
11. What have the emu, the rhea and the ostrich in common?
12. Wallabies and kangaroos are found mainly in which country?
13. What is the tallest living animal?
14. How long can an Emperor penguin stay submerged?
15. How much of the world's ice is contained in the Antarctic continent?

ANSWERS

1   The jaguar.

2   The orang-utan.

3   A slow-moving mammal found among the trees of South America. It feeds mainly on leaves and fruit.

4   The three-banded armadillo found on the South American plains.

5   A large flightless bird.

6   A large, hen-like, swamp bird. Thought extinct in the 1890s they were rediscovered in 1948 and there are just over 180 left in New Zealand.

7   They are birds of prey of which nine species are known.

8   There are sixteen species (18 if one considers the royal and the white-flippered a full species).

9   The African elephant.

10 The bison. It stands 2.2 metres (7 feet 3 inches) high at the shoulder.

11 They are all flightless.

12 Australia.

13 The giraffe. It can measure up to 5 metres (17 feet).

14 They can submerge for 18 minutes and reach depths of at least 265 metres (860 feet).

15 90%.

Wildlife Quiz

1. What is a raptor?
2. Where on a bat would you find its tragus?
3. What is a polder?
4. What are a bird's remiges?
5. What is a red tide?
6. What is the word used for a young duck?
7. How many spines does an adult hedgehog have on its back?
8. A cushat is a popular alternative name for what bird?
9. Where would you find the anther?
10. Where would you find a lateral line?
11. What is photosynthesis?
12. What does the word 'xylophagous' mean?
13. Name two of the three major cloud groups?
14. What is the Antarctic Convergence?
15. What are the food plants of the caterpillar (larva) of an eyed hawk-moth?

ANSWERS

1. A name used indicating bird of prey.
2. Inside the ear. It is a distinctive lobe and can assist in identification.
3. Reclaimed land from the sea which is protected by an embankment.
4. The main flight feathers of a bird (primaries and secondaries).
5. The bloom of a single-celled organism in the sea which often gives a rusty colour to the area of water affected.
6. Duckling.
7. Approximately 12,000.
8. Wood pigeon.
9. In a plant. It is the tip of the stamen which produces the male pollen.
10. The line that runs along the side of a fish, from the gills to the tail, and is full of nerves in order that the fish can sense what is going on around it.
11. The process through which plants make food from sunlight and water. Chlorophyll (the substance that gives the green colour in leaves) absorbs the energy from sunlight.
12. Wood-eating.
13. The three major cloud groups are: cumulus or heap clouds, stratus or sheet clouds and cirrus or fibrous clouds.
14. Situated between 50°S and 60°S, this is where the cold waters of the Antarctic region meet the warmer waters from the middle latitudes (Atlantic, Pacific and Indian Oceans). The cold waters sink beneath the warmer waters. The convergence area is organically rich and teaming with wildlife species.
15. Predominantly willow species including grey, sallow, goat, white, and osier. They also eat apple and occasionally plum, rose and poplar.

Wildlife Quiz

1. What does *Lepidoptera* mean?
2. One of the differences between the common newt and the viviparous lizard is the number of toes on the front foot. Which one has the most?
3. Where are the oldest known footprints situated in Ireland and what type of creature made them?
4. What plant is associated with the legendary werewolf?
5. What is another name for reindeer?
6. How many mammal species are there in the world?
7. How many bird species are there in the world?
8. Ireland has just one breeding species of bird that is globally threatened. Which is it?
9. What bird is known as 'the snipe of the woods'?
10. Which is Ireland and Britain's smallest duck species?
11. What is a greenshank?
12. Where does the reed warbler winter?
13. Which family does the otter belong to?
14. What is the favourite food of the caterpillar of the small tortoiseshell butterfly?
15. How many butterfly species have been recorded in Ireland?

ANSWERS

1. It means scaly-winged and is the scientific name of the order that butterflies and moths belong to in the animal kingdom.
2. The common newt has 4 front toes and the viviparous lizard has 5 front toes.
3. On Valentia Island, Co. Kerry. They were made by a tetrapod, a cross between an amphibian and a reptile, 385 million years ago. That is 150 million years before dinosaurs walked the earth.
4. Wolfbane.
5. Caribou.
6. There are 4,629 known species at present.
7. There are approximately 8,700 species at present.
8. Corncrake.
9. Woodcock.
10. Teal.
11. An attractive long-legged, long-billed, wading bird found in estuaries and along tidal areas.
12. In Africa.
13. The *Mustelidae* family which includes weasels, stoats, mink and martens. The badger is also a member of this family.
14. Stinging nettles.
15. 40 species. This includes six rare migrant vagrants, two species that are now extinct, and one introduced species.

Wildlife Quiz

1 What migrant hawk-moth is named after a small American bird?

2 What is a painted lady?

3 What is a hoopoe?

4 What is an alternative name for a corncrake?

5 Which two common owl species breeding in Britain, do not occur in Ireland? Is it long-eared owl, short-eared owl, tawny owl, barn owl or little owl?

6 Sandwich, common, little, arctic and roseate are the names associated with which bird family?

7 Which bird is found on the Irish 50 pence coin?

8 What animal is found on the Irish punt coin?

9 What small wading bird gets its name from turning pebbles and small stones along the shore line?

10 Why do birds, such as pied wagtails, fly at windows and car mirrors?

11 What is a backswimmer?

12 What is a newt?

13 Do oystercatchers eat oysters?

14 What is the common name for the green plover?

15 What bird is used as the logo for the R.S.P.B. (Royal Society for the Protection of Birds)?

ANSWERS

1. Humming-bird hawk-moth. It gets its name from the way it hovers in front of flowers as it feeds.

2. A migrant butterfly.

3. A bird with an impressive crest that gets its name from its call. Common in parts of Europe but rare in Ireland and Britain.

4. A land rail. Closely related to the water rail.

5. Tawny owl and little owl.

6. The tern family.

7. The woodcock.

8. The red deer.

9. The turnstone.

10. The bird sees its reflection in the window or mirror, thinks it is another bird and wants to fight over territorial rights.

11. A 'true bug' that swims upside down in ponds and streams. It uses its back legs as oars and, when diving, traps an air-bubble against its abdomen.

12. An amphibian (frogs, toads, salamanders and newts). They prefer damp environments.

13. No. Their diet consists mainly of limpets, mussels, worms and other marine invertebrates.

14. The lapwing or pewit.

15. The avocet.

Wildlife Quiz

1. What bird is used as the logo for BirdWatch Ireland?
2. Which famous English artist did the drawing for the BirdWatch Ireland logo?
3. What is a ring ouzel?
4. Which two species of thrush, not present in the summer, are regular winter migrants to Ireland?
5. What is a dread?
6. What is a mammal?
7. What is a carnivore?
8. What is a female red fox called?
9. What species of mammal did the Normans introduce to Ireland?
10. When was the grey squirrel introduced into Ireland?
11. What percentage of water on the earth is fresh?
12. Name the two main types of bogland found in Ireland?
13. Pond skater, whirligig beetle, flatworm, mosquito larva, water beetle and springtail can be found mainly where?
14. Name the four stages of a butterfly's development?
15. What is the Irish name for a bat?

ANSWERS

1 The Greenland white-fronted goose.

2 The late Sir Peter Scott.

3 A blackbird-like thrush. The male has a white crescent-shaped breast band. They breed in mountainous areas and are a scarce summer migrant to Ireland.

4 The redwing and the fieldfare.

5 A sudden apparent wave of alarm that spreads through a flock of birds (particularly waders, gulls and terns).

6 A warm-blooded creature with a bony internal skeleton that suckles its young.

7 An animal or plant which feeds upon other animals.

8 A vixen.

9 The rabbit, which was introduced 800 years ago.

10 In 1911. They were introduced into Co. Longford.

11 3%.

12 Raised bogs and mountain blanket bogs.

13 On the surface of a pond.

14 1. egg, 2. caterpillar (larva), 3. chrysalis (pupa), and 4. butterfly (adult).

15 Ialtóg.

Wildlife Quiz

1. What fish part, often found on the seashore, is given to caged birds?
2. What does 'pelagic' mean?
3. Where is the hadal zone?
4. What is a tsunami?
5. Where does the Gulf Stream begin?
6. How many eggs does a female lobster release into the sea?
7. What is a female seal called?
8. What are the two species of seal found in Irish waters?
9. What is a frugivore?
10. What seal is named after a musical instrument?
11. 1% of the ocean surface produces 60% of the world's marine fish catches. True or false?
12. What is a baleen whale?
13. A whale's tail is horizontal and moves up and down, whereas a fish's tail is vertical and moves from side to side. True or false?
14. What is the name of the fin on the top of the whale's back?
15. Where would you normally find the grey whale?

ANSWERS

1   Cuttlefish. The backbone is used by pet birds to sharpen their bills.

2   Of the open sea. It applies to organisms that inhabit open water and also applies to birds that only come to land to breed.

3   The part of the ocean that lies in very deep trenches below the general level of the ocean floor.

4   A tidal wave.

5   It begins near the coast of Florida.

6   Between 5,000 and 40,000.

7   A cow.

8   The common, and the Atlantic grey.

9   A seed or fruit eater.

10 The harp seal.

11 True.

12 A filter feeding whale.

13 True. At a glance it is the easiest way to tell the difference between them.

14 The dorsal fin.

15 The North Pacific and in the Arctic oceans.

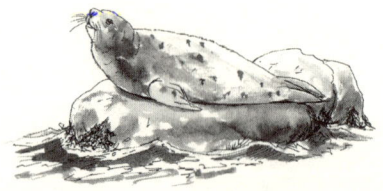

Wildlife Quiz

1. What is another name for a killer whale?
2. What tree is known as 'the lady of the woods'?
3. What seabird has, in recent years, colonised rooftops in towns and cities?
4. In Herman Melville's novel *Moby Dick*, what species was the white whale?
5. What is an omnivore?
6. What species of dolphin is 'Fungie' in Dingle, Co. Kerry?
7. What is a parasite?
8. What is a predator?
9. What is a toad-fly?
10. What are the other names given to the mountain lion of North America?
11. Where would you find the snow leopard?
12. In America what animal is called the song dog?
13. What is a chinchilla?
14. What is an argali?
15. Why do seals appear to cry when on land?

ANSWERS

1. An orca.
2. Silver birch.
3. The herring gull.
4. Sperm whale.
5. An animal that eats plants and animals.
6. A bottlenose dolphin.
7. An animal or plant that lives in or on another to obtain its nourishment.
8. An animal that feeds upon other animals.
9. A fly which lays its eggs on an unsuspecting toad. On hatching the maggots travel, internally, to the toads head and slowly eat it from within.
10. The puma and the cougar.
11. In the mountains of central Asia.
12. The coyote.
13. A rodent living in the Andes in South America.
14. A member of the sheep family with heavy horns. The animal is highly agile and found in central Asia.
15. Seals have no tear ducts and as a result the fluid spills out on to their fur. Normally this would be washed away by the water they live in.

Amazing Facts

Animal Imprinting
When a zebra foal is born, its mother prevents it from looking at any other zebra until it becomes familiar with her body pattern. This imprinting may take up to twenty minutes. From then onwards, the foal will always recognise its mother's pattern.

Snakes Alive
40,000 people die of snake bites each year. 75% of those deaths occur in India alone.

At Arms Length
One of the most deadly creatures in the sea is the blue-ringed octopus. Its venomous bite can kill a man within 2 hours.

Fisher Kings
Some fishermen in Japan use cormorants to catch fish for them. They tie a collar around the bird's neck to prevent the bird swallowing the catch.

Busy Queen
A termite queen can lay up to 1,000 eggs a day at a rate of one per minute.

Leaf Legend
The welwitschia plant of Namibia lives on dew and has leaves that grow up to 18 metres (58 feet) long.

Soak It Up
Sphagnum moss, found in bogs, can soak up to 25 times its own dry weight.

Bird Quiz

1. What is the smallest bird in the world?
2. What is the world's fastest bird?
3. Which bird became a celebrity when it nested for the first time in 1959 at Loch Garten, Scotland?
4. What species of bird has the greatest wingspan in the world?
5. What is the greatest wing-span of any landbird?
6. What is the world's heaviest bird?
7. What is the world's heaviest flying bird?
8. What is Ireland and Britain's heaviest flying bird?
9. What is the only bird species in the world that has a sideways-curving bill?
10. What bird has the most feathers?
11. What bird has the least amount of feathers?
12. What is the only entirely flightless parrot in the world?
13. What is the highest record of a bird in flight?
14. What species of bird has the fastest wingbeats?
15. What is a galah?

ANSWERS

1. The mole bee hummingbird from Cuba and the Isle of Pines, is just 5.7 cm. (2.24 inches) in size. A male weighs just 1.6 gms.
2. The peregrine falcon has been recorded at up to 180 kp/h. (112.5 mph.) in a dive. The male is believed to travel much faster when in display flights with speeds of up to 360 km/h. (225 mph.) claimed, but there are questions as to the accuracy of these figures.
3. The osprey.
4. The wandering albatross with a wingspan of up to 3.63 metres (11 feet 11 inches).
5. The Andean condor and the marabou stork both have wingspans of up to 3.2 metres (10 feet 5 inches).
6. The ostrich, which is also the tallest bird in the world. The heaviest and tallest birds are of the North African subspecies and have been recorded as weighing 156 kilograms (345 lbs.) and up to 2.7 metres (9 feet) high.
7. In the wild, the great bustard is credited as being the heaviest species to regularly fly. Adult males often reach 16.8 kilograms (37 lbs.) with extremes recorded at 18 to 21 kilograms (39.7 to 46.3 lbs.).
8. The mute swan, with the male often weighing up to 14 kilograms (31 lbs.).
9. The wrybill, which is a species found in New Zealand. The end quarter of its bill turns to the right by 12°. It uses its bill to probe for insects under stones.
10. The whistling swan which has 25,216. 80% (over 20,000) of these were found on the head and neck of the swan.
11. The ruby-throated hummingbird with just 940 feathers.
12. The kakapo, which is found in New Zealand.
13. A Ruppell's griffon vulture was recorded at 11,274 m. (37,000 feet) on 29th November 1973 when it collided with a commercial aircraft over the Ivory Coast, Africa. The plane landed safely though the engine had to be shut down. The bird was identified from feathers in the engine.
14. The amethyst woodstar hummingbird has had its wingbeats recorded at 90 beats a second. Faster speeds have been recorded but these were for the tips of the flight feathers as opposed to the whole wing.
15. An Australian cockatoo. This species was formally known as the roseate cockatoo.

Wildlife Quiz

1. What is a garden tiger?
2. What is the 'cock-of-the-rock?
3. Where would you find the crown eagle?
4. What is the ceiling of the jungle called?
5. Where do hummingbirds live?
6. Where do sunbirds live?
7. Fruit bats cannot perch on branches nor use their 'arms' to pluck fruit, so how do they feed?
8. What animal produces the loudest sounds in the animal kingdom?
9. What is a tamandua?
10. What is an elephant shrew?
11. What is particularly unusual about some species of jungle mushroom?
12. What is the average thickness of the ice on the Antarctic continent?
13. Who was the first person to cross the Antarctic circle in 1773?
14. What is the largest desert in the world?
15. What is a thorny devil?

ANSWERS

1. A common moth found in Ireland and Britain.
2. A bird of paradise found in South America.
3. In Africa.
4. The canopy.
5. In America.
6. In Africa.
7. They flop onto a branch, grab the fruit in their mouths and fly off.
8. The howler monkey.
9. A South American anteater.
10. A rodent, the size of a cat, found in Africa. It gathers food using its trunk-like snout.
11. They can glow in the dark.
12. 2,000 metres.
13. Captain James Cook.
14. The Sahara which is almost 8.5 million square kilometres (5.3 million square miles) in size.
15. A spiny lizard found in the deserts of central Australia.

Wildlife Quiz

1. What is a jack rabbit?
2. What is a road runner?
3. How many eyes does a spider have?
4. What is particularly unusual about the timing of the Emperor penguin's nesting season?
5. What is an xenops?
6. What is a tui?
7. What is the fastest-running bird?
8. When and where was the collared dove first seen in Britain?
9. When and where was the collared dove first seen in Ireland?
10. In 'Who killed Cock Robin' what did the owl say?
11. What colour is the flower of the sea campion?
12. Does a juvenile robin have a red breast, orange breast, or a brown speckled breast?
13. What is a chickadee?
14. Where does the giant otter live?
15. What is the first bird mentioned in the bible?

ANSWERS

1. A hare found in California and other parts of southern America.
2. A bird that lives in desert regions of America and is a member of the cuckoo family.
3. Most species have eight. There are nine species that have six.
4. One egg is laid during the Antarctic winter in temperatures around −60° centigrade. The male, who broods the egg, has to fast for between 110 and 115 days during this period.
5. There are five species of xenops, members of the ovenbird family.
6. A honeyeater bird found in New Zealand and off-lying islands.
7. The ostrich can reach speeds of up to 96.5 km/h. (60.3 mph.).
8. On 31st July 1952 at Manton in Lincolnshire.
9. In 1959 in a Dublin suburb.
10. "I'll dig his grave".
11. White.
12. A brown speckled breast.
13. A small agile member of the tit family found in North America.
14. Along the Amazon River.
15. Raven.

Wildlife Quiz

1. In *Treasure Island*, what did Long John Silver's parrot say?
2. What colour is the flower of a lesser hawkbit?
3. What does the word 'jizz' mean?
4. What species of bird is also known by the local name 'stare'?
5. What scale is used to measure windspeed?
6. What fish found in Ireland is regarded as the freshwater equivalent of a shark?
7. What is a gatekeeper?
8. Gavin Maxwell's book *A Ring of Bright Water*, which was made into a film, featured which animal?
9. The marsh tit, a member of the tit family, is fairly common in Ireland. True or false?
10. Can you name Ireland's only species of wild dog?
11. What are 'lords and ladies'?
12. If a monkey's tail is described as prehensile, what does it mean?
13. What does a humpback whale eat?
14. What is the largest reptile found in the United States?
15. What is unusual about the baobab tree of Africa?

ANSWERS

1 "Pieces of eight".

2 Yellow.

3 A number of features that identify a living creature in the field. e.g. The way it may move or fly, how it catches its prey etc.

4 The starling.

5 The Beaufort Scale.

6 The pike.

7 One of the brown butterflies.

8 The otter.

9 False.

10 The fox.

11 These plants, also known as 'wild arum', 'cuckoo-pint' and 'parson-in-the-pulpit', produce a stem covered in orange-red berries in autumn.

12 Capable of grasping.

13 Krill (very small shrimp-like animals).

14 The alligator [specimens of up to six metres (20 feet) have been recorded].

15 It grows upside-down.

Wildlife Quiz

1. What is the name of a squirrel's home?
2. What is a 'ha ha'?
3. What is sometimes called a 'lawyer's wig'?
4. What are ruminants?
5. How old is the sun?
6. What is a wainscot?
7. What is the common name for the lesser rorqual?
8. What is a cockchafer?
9. What Walt Disney wildlife film won an Academy Award?
10. What are catchflies?
11. The Egyptian goose breeds in Africa south of the Sahara. Where else does it breed?
12. Grasshoppers and crickets belong to what family group.
13. Sometimes a stag hides in very thick cover. What is this lair called?
14. Where is the largest colony of great crested grebes in Ireland?
15. The 'mad man's window' is the name of a strange rock formation. Where is it situated?

ANSWERS

1. Drey.

2. An old ditch, sunken fence, or boundary wall designed to keep deer and cattle within the grounds of an estate, without breaking the sky line.

3. Another name for the shaggy cap fungus (an edible fungus).

4. They are cud-chewing animals and are herbivores (they have a complex four stage stomach, one part is a store where food is held pending mastication – chewing the cud).

5. 4,600 million years old.

6. A moth. There are a number of different species of wainscot recorded in Ireland and Britain.

7. The minke whale.

8. A large beetle often referred to as the May bug. They can be very destructive to crops.

9. *The Living Desert.*

10. A perennial plant that has a number of species throughout Ireland and Britain.

11. Amongst other places, there is a feral flock breeding and wintering in Norfolk, England.

12. *Orthoptera.*

13. A harbour.

14. Lough Neagh, Co. Antrim.

15. Glenarm, Co. Down.

Wildlife Quiz

1. Redshank is the name of a common bird and common plant. True or false?
2. How many earwig species are there in Europe?
3. Where do budgerigars come from?
4. What is the name of a badger's home?
5. What colour is the flower of a spring gentian?
6. What magpie is not a bird?
7. What is a peacock? Is it a bird, a butterfly or a worm?
8. Which is Ireland's largest gull species?
9. What is adderstongue?
10. What is a chelae?
11. Where would you find the Beagle Channel?
12. What is a baby kangaroo called?
13. What is a redpoll?
14. What is wall rue?
15. What toad species can be found in Ireland?

ANSWERS

1. True. Redshank is a common wading bird and redshank is also a plant found on waste ground.
2. There are 34 species of earwig in Europe of which only 4 have been recorded in Britain. World-wide there are 1,300 different species.
3. From Australia.
4. A sett.
5. Bright blue.
6. The magpie moth.
7. All three answers are correct.
8. The great black-backed gull.
9. A fern-type plant found in dune slacks, damp ground and old meadows.
10. The pincers of crabs and lobsters.
11. Tierra del Fuego between Argentina and Chile. Named after the ship of Charles Darwin who formulated the theory of evolution.
12. A joey.
13. A small bird belonging to the finch family.
14. A small fern, which is found growing on walls.
15. The natterjack, which is present in Co. Kerry and Co. Wexford.

Wildlife Quiz

1. What is a kookaburra?
2. What two separate organisms combine to form a lichen?
3. What is a crinan ear?
4. The *Aminita muscaria* can often be found in illustrations in fairy tale books. What is it?
5. What colour are starlings' eggs?
6. What does 'vestigial' mean?
7. What is a 'mind-your-own-business'?
8. What is a harlequin?
9. What burrowing mammal, the bane of English gardeners, is not present in Ireland?
10. What is a coarse fish?
11. The remains of an elephant-like mammal, now extinct, have been found in Ireland. What is it?
12. What colour are shelduck chicks?
13. What is phycology?
14. What is the name given to the seed of the beech tree?
15. What animal was traditionally used to find truffles?

ANSWERS

1   A bird of the kingfisher family native to Australia.

2   Certain types of algae and fungus create a symbiotic relationship and form lichens.

3   A *Noctuidae* moth.

4   The fly agaric, a fungus with usually a red cap covered in white spots.

5   Blue.

6   Part of an animal that tends to be small, serves no useful purpose, and is disappearing through the evolutionary process.

7   A small creeping perennial herb.

8   A North American duck.

9   The mole.

10 Any freshwater fish not of the salmon family.

11 Woolly mammoth.

12 Black and white.

13 The study of algae.

14 Beech mast.

15 The pig (the truffle is an underground fruiting body of a fungus).

Wildlife Quiz

1. What does 'crepuscular' mean?
2. What is a the name of the shell that covers a turtle's body?
3. What does a conchologist study?
4. What colour is the flower of the birdsfoot trefoil?
5. Which group of insects has the largest eyes?
6. What part of a snail shell is called the peristome?
7. What is a hamerkop?
8. The *Arbutus unedo* is the scientific name for a native tree found in south-west Ireland and southern Europe. What is its common name?
9. In which countries would you find trapdoor spiders?
10. Where does a goldeneye duck make its nest?
11. There is only one shrew species in Ireland. What is it?
12. Where would you find a chamomile shark?
13. The male sea horse gives birth to its live young. True or false?
14. How many eggs does an ostrich lay?
15. What are marmosets?

ANSWERS

1. It literally means 'of the twilight' and is applied to organisms that are active at dawn and dusk.
2. The carapace.
3. Molluscs. The word means 'studiers of shell fish' but it includes slugs and snails.
4. Yellow, often tinged orange or red.
5. The dragonfly family. The largest-eyed of these belongs to a group of hawker dragonflies that live in shady tropical rainforests and hunt during the dawn and dusk periods – thus needing those particularly large eyes.
6. The mouth and mouth edge of the shell.
7. A bird with alternative names of hammer-headed stork and anvil-head. It is found throughout the tropical regions of Africa, Madagascar and south-west Arabia.
8. The strawberry tree.
9. In Africa, the Americas, and Australia. They live in silk-lined tubes dug into the ground and covered by silk lids. Prey are attacked and pulled into the tube.
10. In the natural cavity of a tree or stump. Woodpecker holes are also used.
11. The pygmy shrew, the scientific name being *Sorex minutus*.
12. Not in the sea, as you might expect, but on fence posts. Believe it or not, it is a moth.
13. True. The male gives birth to about 1,000 young from his stomach pouch.
14. Normally about 12. Other females may use the same nest and as many as 40 eggs can be found in a single nest.
15. They are tiny monkeys that live in the canopy of the South American rainforests.

Amazing Facts

Snail Tale
The eggs of the common garden snail are used in some laboratories to help determine human blood groups.

Swift Flight
A common swift has one of the most amazing first flight records. The maiden voyage of a young swift leaving its nest can last up to three years. Swifts are adapted to eat, drink and sleep on the wing. Their whole life is adapted to flight. One of the most amazing bird-ringing recoveries was of a swift ringed in Oxford as a nestling and recovered near Madrid three days after fledging. That is a distance of 1,300 km. (813 miles) and the bird was just out of the nest!

Mother's Care
The female sand grouse transports water to its young by soaking its breast and allowing the young to drink the water from its feathers.

Heavy Weight
A pinhead-sized piece of a neutron star weighs one million tonnes.

Pigeon Hero
One of the heroes decorated in the First World War was a pigeon, which carried vital messages through artillery fire round Verdun in 1916. The bird died after a hazardous trip, and was awarded a posthumous Legion d'Honneur for its courage.

Bird Ingenuity
The Japanese green heron actually uses live insects as bait when fishing for food – just like a human fly-fishing. An amazing example of tool-using in the animal kingdom.

Wild Flower Quiz

1. What special trait of the alpine snowbell enables it to flower in early spring?
2. What is the fastest-growing plant in the world?
3. What is anemophily?
4. What type of plant is a bristly ox-tongue?
5. What is a monk's-hood?
6. What plants are useful indicators of air pollution?
7. What is transpiration?
8. Cork is obtained from what kind of tree?
9. What colour is the fruit of the guelder rose?
10. What is a samara?
11. Angiosperms is the collective name for what?
12. What plant produces the largest seed in the world?
13. What water plant is said to produce some of the most poisonous natural substances?
14. How does the sphagnum moss disperse its spores?
15. How strong are the leaves of the giant South American water lily?

ANSWERS

1. It actually generates sufficient heat to melt a hole in the snow.
2. The bamboo. Some species grow up to 91 cm. (3 feet) a day.
3. Pollination of plants by wind.
4. A thistle.
5. A poisonous, purple-flowered plant that is a member of the buttercup family. In ancient times hunters used an extract from its roots on the tips of their arrows.
6. Lichens. Most species cannot tolerate pollution.
7. The process by which plants lose water vapour, through leaves and stems.
8. The cork oak tree, found in Southern Europe and North Africa.
9. Red.
10. A winged key-shaped fruit.
11. The flowering plants. Non-flowering plants are known as gymnosperms.
12. The giant fan palm, also called the coco-de-mer. It grows exclusively in the Seychelles and produces a single-seeded fruit that can weigh up to 20 kg. (44 lbs.) and can take up to ten years to develop.
13. Red tide algae. The toxins produced by these microscopic plants have been known to poison humans. Research on red tide algae is being carried out in Ireland.
14. The seed capsule, as it ripens, contracts to a quarter of its size, compressing the air within to double the pressure of a car tyre. When the time is right the lid blows off with a pop and the spores are fired out as if from a gun.
15. This plant's huge leaves, 2 metres (7 feet) across, which float upon the water, can support the weight of a child.

Nature Quiz

1. How does an egg-eating snake swallow a large egg?
2. What is a pheromone?
3. What is the colour of the flower of a meadow vetchling?
4. What is the other common name for the cross spider?
5. What is the fastest animal on four feet?
6. What is a san orca?
7. Where do puffins nest?
8. What is rookooing?
9. Name the two chipmunks often featured in a Donald Duck cartoon?
10. What is another name for the woodbine?
11. What does 'calcareous' mean?
12. The yellowhammer is a common bird of farmland and open country. What is another name for it?
13. What beetle sounds like it has just been in a fight?
14. How many eggs does a female guillemot lay in a season?
15. Do birds mate for life?

ANSWERS

1. It dislocates its lower jaw from the upper jaw to enable swallowing. It then pushes a sharp back bone forward to puncture the egg shell.
2. A chemical scent produced by an organism and released into the surrounding area, which then causes a response in another individual of the same species.
3. Yellow.
4. The garden spider.
5. The cheetah.
6. A false killer whale.
7. In burrows on steep grassy slopes.
8. The production of a bubbling sound by the male black grouse at a lek – the communal display ground.
9. Chip and Dale.
10. Honeysuckle.
11. Made of chalk or calcium carbonate.
12. Yellow bunting.
13. The bloody-nosed beetle. A flightless beetle which, when disturbed, produces a bright red fluid from its mouth – hence its name.
14. Just one large egg which is laid on a cliff-ledge.
15. Most birds probably do. However small birds do not live long so it is the larger ones, like swans, which get the benefit of a reputation for being faithful.

Nature Quiz

1. Where would you find witches' butter and what is it?
2. How many families of baleen whale are there?
3. The Abingdon Island giant tortoise is an endangered species. How many individuals are there left in the world?
4. The clouded-bordered brindle moth larvae eats what common spring-flowering plant?
5. Where would you find the alula on a bird?
6. What colour is the flower of the common butterwort?
7. What does 'neritic' indicate?
8. The scientific name *Puffinus puffinus* relates to what bird?
9. Can you name the four species of great ape which are man's closest living relatives?
10. Whose underwater garden did the Beatles want to be in?
11. What is 'the confused'?
12. What was the name of the tiger in Rudyard Kipling's *The Jungle Book*?
13. When threatened some lizards can shed their tails. True or false?
14. How does an archerfish catch its prey?
15. What colour are yew tree berries when ripe?

ANSWERS

1. Witches' butter is a common fungus found on deciduous trees.
2. Three families. There is one species of grey whale; three species of right whale and six species of rorqual whale.
3. Just one, a male called Lonesome George. When he expires the species becomes extinct.
4. The primrose and the cowslip.
5. The small tract of feather attached to the first digit of the wing.
6. Violet.
7. The term relates to the near shore, the shallow water zone of the sea over the continental shelf.
8. The manx shearwater.
9. Chimpanzee, gorilla, orang-utan and the bonobo.
10. In an octopus's garden.
11. A moth of local distribution in Ireland.
12. Shere Khan.
13. True. The process is called autotomy.
14. It squirts jets of water from its mouth to knock insects from overhanging vegetation.
15. Scarlet.

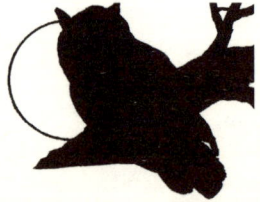

Nature Quiz

1. A barn owl's vision is superior to a human's. Is it 10 times better, 50 times better or 100 times better?
2. What name is the great reed mace better known as?
3. What is the name given to the Australian wild dog?
4. What is the sloe the fruit of?
5. What species of moth featured in the film *The Silence of the Lambs*?
6. What is the main physical difference between an African and an Indian elephant?
7. What is a pochard?
8. What plant is the national emblem of Scotland?
9. What colour is a jackdaw's eye?
10. What birds are known as 'Mother Carey's Chickens'?
11. How many vertebrae do mammals have in their neck?
12. Certain fungi feed on small worms called nematodes. How do they catch them?
13. What is a motmot?
14. The sycamore tree is not native to Ireland. Who introduced it?
15. In Samson's riddle, where was the honey found?

ANSWERS

1. 100 times better. That is bad news if you are a mouse!
2. The bulrush or cattail.
3. A dingo.
4. The blackthorn.
5. The death's-head hawk-moth.
6. The African elephant has larger ears than the Indian elephant.
7. A common diving duck.
8. The thistle.
9. Light blue.
10. Various species of storm petrel.
11. Seven. All mammals have the same number of vertebrae in their necks.
12. Each fungus grows a 'noose' and as a worm is detected crawling through, the noose tightens instantly, the worm is trapped and slowly devoured.
13. Motmots are a group of birds quite closely related to kingfishers.
14. It was introduced by the Romans.
15. In the body of a dead lion.

Nature Quiz

1. What plant was traditionally used for making rope?
2. What is the thallus?
3. What colour is the bill of an adult male blackbird?
4. Where does an adult pond skater go in the winter?
5. What colour is the sexton beetle?
6. Where is mare's tail found?
7. The glow worm is a small worm. True or false?
8. What is Chinese bird's-nest soup made from?
9. Longhorn beetles are a family of more than 20,000 species. What do the larvae of these beetles eat?
10. What porpoise has a name which suggests that its eyesight is poor?
11. What does the death cap, the destroying angel and the fool's mushroom have in common?
12. What is particularly distinctive about the appearance of the banded demoiselle?
13. What is a spinney?
14. What colour are the legs of a black-winged stilt?
15. In Ireland the hare, stoat, red grouse, dipper, jay and coal tit all have something in common. What is it?

ANSWERS

1. Hemp.
2. It is the main body of a lichen, and does not have leaves, stem or roots.
3. Yellow.
4. They fly away from water and hibernate.
5. Black with large orange-red markings on the back.
6. This common freshwater plant is found in lakes and ponds.
7. False. The glow worm is a beetle. Only the male has wings.
8. The nests of cave swiftlets which are made from their hardened saliva.
9. They are almost all wood-eating, attacking both living and dead timber.
10. The spectacled porpoise.
11. They are all deadly poisonous fungi. The death cap accounts for 90% of all deaths attributed to fungi, the other two being much rarer species.
12. This damselfly has an iridescent green or blue body (depending on whether male or female) and the transparent wings have a dark band running across the centre from front to rear.
13. A small wood.
14. This bird has very long red legs.
15. They are all Irish sub-species.

Nature Quiz

1. What is the name of a fox's home?
2. What are the rectrices of a bird?
3. The eclipse of the sun on August the 11th 1999 is part of Saros series 145. When did this series start?
4. What comes between the earth and the sun to cause a solar eclipse? Is it the moon, a comet or a meteor shower?
5. What is a John Dory?
6. What bird featured in the film *Kes*?
7. What is tormentil?
8. What is the butterfly bush more commonly known as?
9. What colour are the under-tail feathers of a grey wagtail?
10. What are *Arachnida*?
11. What is a tucuxi?
12. What is a nutcracker?
13. Where and how big was the largest earthworm ever recorded in the world?
14. What common bird species used to be known in a variety of areas as a 'bum barrel'?
15. What colour are the flowers of houndstongue?

ANSWERS

1 Den or earth.

2 The main tail feathers.

3 1639.

4 The moon.

5 A sea fish.

6 A kestrel.

7 A widespread and abundant perennial plant found in grassy places, moors and bogs.

8 Buddleia, which comes from its scientific name – *Buddleia davidii*.

9 A bright lemon-yellow. Because of this they are sometimes mistaken for a yellow wagtail.

10 Spiders which include mites, ticks, scorpions and pseudoscorpions.

11 This is one of the smallest dolphins and is also known as the estuarine dolphin.

12 A bird belonging to the *Corvidae* (crow) family. Common over Europe but a rare visitor to Britain.

13 The specimen was found in South Africa in 1937 and measured 6.7 metres long (21.8 feet) and was 2 cm. (1 inch) in diameter.

14 The long-tailed tit.

15 Maroon or darkish red.

Nature Quiz

1. Where would you find an insect called a large heath?
2. What does the word 'dehiscent' mean?
3. How many broods of eggs does a swallow normally lay in a season?
4. What is a chukar?
5. Where would you find a chiton?
6. What is a rhizome?
7. How many beaked whale species have been identified to date?
8. What is the difference between an entomologist and an etymologist?
9. What is the technical name for woodlice?
10. How many known species of ant are there in the world. Is it 10,000; 15,000; or 20,000?
11. The Noble fir tree is renowned for the large size of its cones. How many seeds does a cone hold?
12. What does 'precocial' mean?
13. What is a reeve?
14. What colour is a cinnabar caterpillar?
15. What is a cowrie?

ANSWERS

1. The large heath is a butterfly found in bogland areas.
2. It means bursting or splitting open at maturity and is normally associated with a seed pod bursting open to release seeds.
3. They are normally double or treble brooded.
4. A bird that is one of the partridge family and found in south-eastern Europe. The species is currently declining.
5. This mollusc is found attached to rocks on rocky shores where it grazes on tiny seaweeds.
6. A horizontal underground stem which bears leaf scars.
7. 20 species to date with some species never having been seen alive.
8. An entomologist is a person who studies insects whereas an etymologist is a person who studies the history and origin of words.
9. Isopoda.
10. 15,000 known species.
11. Up to 1,000.
12. It is another word for nidifugous which is a chick that hatches with down on and is able to leave the nest shortly after hatching.
13. The female of a ruff, which is a wading bird.
14. Orange and black.
15. A mollusc whose shell has a slit-like opening on its lower side. Small examples are found on our western coasts with larger species found in tropical waters.

Nature Quiz

1. What is a robin's pincushion?
2. What is a moschatel?
3. What plant is the national emblem of Ireland?
4. The spurge family of plants produce a milky substance from their stems. Why?
5. What is royal jelly?
6. What is common sorrel?
7. What is a lammergeier?
8. 'Who killed cock robin' in the nursery rhyme?
9. What are barracuda?
10. What colour are the flowers of mallow?
11. How many species of mole are there world-wide?
12. What is the largest living amphibian in the world?
13. Where would you find thongweed?
14. What are palps?
15. What is an awn?

ANSWERS

1. A fibrous growth on the dog rose which is caused by the larvae of one of the gall wasps. Initially green it has a reddish tint in the autumn.
2. A perennial pink-flowering plant of woodland and shady places.
3. The shamrock.
4. This natural defence mechanism is created by the plant to prevent insects eating the foliage. When an insect tries, its mouth parts get gummed up by the sticky substance.
5. The food secreted and fed by worker bees to developing queen bees.
6. An edible perennial plant that has a sharp taste.
7. A very large vulture also known as the bearded vulture.
8. The sparrow.
9. Predatory fish living in shoals in the open sea.
10. Pink.
11. There are 27 species identified to date.
12. The Japanese giant salamander which grows up to 1.5 metres (5 feet) long and can weigh up to 100 kg. (220 lbs.).
13. On the lower seashore. It is a common seaweed that attaches itself to rocks.
14. Sensory organs found around the mouths of insects and crustaceans.
15. A bristle found in the flowers of many grasses.

Amazing Facts

Song Of The Birds
All birds use calls to communicate. Some birds repeat their song over 1,000 times a day.

The Earth?
Almost 71% of the earth's surface is covered by oceans.

Wet Baby
Our youngest ocean is the Atlantic. It was formed 200 million years ago by the surrounding continents splitting apart.

Making A Splash
The Pacific Ocean is the world's largest ocean. It covers over 33% of the world's surface.

Hidden Depths
The deepest part of the ocean floor is 11,022 metres (36,160 feet).

Killer Wave
Tsunamis are tidal waves which are not caused by wind, and travel at very high speeds of around 750 km/h. (470 mph.).

Pull From Above
The sun and the moon create the gravitational pull that causes tides. There is an amazing difference in the range of levels varying from as little as under a metre (3 feet) in the Mediterranean to as much as 14.5 metres (47 feet) in The Bay of Fundy, Canada.

Big Meal
The anaconda, one of the largest snakes in the world, can capture and swallow prey as big as an antelope.

Our Seas Quiz

1. What family does the sea-horse belong to?
2. What does 'arthropod' mean?
3. In what sea does the European eel's life begin?
4. What is the largest fish in Irish waters?
5. The monkfish is a member of what fish family?
6. How many tentacles does an octopus have?
7. How many arms and tentacles does a squid have?
8. What is a smooth hound?
9. What is a sturgeon?
10. Turbot, brill, halibut, sole and plaice are members of which family?
11. What does a basking shark feed on?
12. What is the largest sea turtle in the world?
13. What is the largest fish in the sea?
14. Which shell does a hermit crab prefer for its protection?
15. What is a mermaid's purse?

ANSWERS

1 The pipefishes.
2 'Jointed-legged'. They include lobsters, crabs, shrimps, etc.
3 In the Sargasso Sea.
4 The basking shark.
5 The shark family.
6 Eight.
7 Eight arms and two long tentacles.
8 A shark.
9 A fish. There are twenty-six species world-wide.
10 Flat fish.
11 Plankton.
12 The leathery turtle.
13 The whale shark.
14 A whelk.
15 The egg case of a dog fish.

Nature Quiz

1. What is a hybrid?
2. What is the main difference between fungi and green plants?
3. Can you name at least two Irish bird species that get their common name from the call they make?
4. What is a planarian?
5. Why is colour important to cuttlefish?
6. What is the name of the Irish organisation interested in wildbirds?
7. What bird features in the opening sequence of R.T.E's *The Late Late Show*?
8. How does the Nile mouthbrooder care for its recently-hatched young?
9. What is the pelage?
10. What bird was featured as the devil's emissary in the movie *Omen II*?
11. What shape is a campanulate flower?
12. Which is larger: the grey squirrel or the red squirrel?
13. In what type of rock are most Irish caves found?
14. What is a radula?
15. What is a Joshua tree?

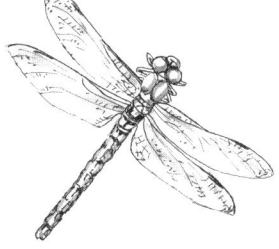

ANSWERS

1. The offspring of two different species.
2. Fungi have no chlorophyll.
3. Chough, chiffchaff, cuckoo, curlew (hoopoe, a rare visitor, can be included).
4. A flatworm.
5. Because of their amazing ability to rapidly change skin colour. Individuals can produce a fantastic variety of colour changes.
6. BirdWatch Ireland, Ruttledge House, 8 Longford Place, Monkstown, Co. Dublin.
7. The barn owl.
8. This fish carries its young in its mouth.
9. The hairy coat of a mammal.
10. The raven.
11. It is bell-shaped.
12. The grey squirrel.
13. Limestone.
14. A snail's tongue.
15. A giant American cactus.

Nature Quiz

1. What is yolk?
2. What does the word 'flora' mean?
3. What does the word 'cryptozoic' in relation to animals mean?
4. What snake is mainly used by snake charmers in India?
5. Burrow duck is the Scottish name for which duck?
6. What is the Bonn Convention?
7. What is another name for the black rat?
8. Ireland holds important populations of two butterfly species that are threatened on a European scale. What are they?
9. What is a rat-king?
10. What does 'alate' mean?
11. What does a herbicide do?
12. A bluebell wood looks beautiful yet was once feared by certain people. Why?
13. What colour is the bill of an arctic tern?
14. What is the stamen of a plant?
15. What is the steep dive of a peregrine falcon called?

ANSWERS

1. The store of food in the egg of the majority of animals (made up of protein and fat granules).

2. It means the vegetation of a place or area.

3. The habitation of crevices such as under stones, logs, leaves etc.

4. The cobra.

5. The shelduck (breeds in burrows).

6. Drawn up at Bonn in 1979 it set out to protect endangered or threatened species of fauna which migrate between different countries. Ireland ratified it in 1983. A number of species are protected by this designation.

7. The ship's rat. Now very rare in Ireland.

8. The marsh fritillary and the large heath.

9. A freak of nature when up to ten rats may be born all tied together by their tails.

10. Winged or having attachments that resemble wings.

11. It kills plants.

12. The Scottish name for the bluebell is 'deadmen's-bells' and superstitious people once believed that to hear the ring of a bluebell was to hear one's death-knell.

13. Red.

14. The male organ of a flower.

15. A stoop.

Nature Quiz

1. What is a tepal?
2. Hector's dolphin is endangered and one of the smallest in the world. They are only found in one area. Where?
3. John James Audubon was one of the great ornithological artists of the 19th century. The famous Audubon Society in America is named after him. Where and in what year was he born?
4. What speed has a golden eagle been recorded at in a vertical dive?
5. What is the head of a tapeworm called?
6. Whales suffer from lice. True or false?
7. What does the 'pecking order' mean?
8. What does 'migration' mean?
9. What is the smallest mammal in the world?
10. What group of mammals are collectively known as pinnipedia?
11. What is an instar?
12. Which is the biggest internal organ found in vertebrates?
13. Which species of seal is the smallest in the world?
14. What animal produces the loudest sound in the world?
15. What function does the haltere on a fly perform?

ANSWERS

1. A tepal is the sepals and petals of rush flowers. The two are indistinguishable – hence the name.
2. Around the coast of New Zealand. They are just 1.4 metres in length (4 feet 9 inches).
3. In Santo Domingo (now the Dominican Republic) in 1785. He was the son of a French father and a Creole mother.
4. 240+ kph. (150+ mph.).
5. The scolex.
6. True. Some species of whale have small crab-like parasites known as whale-lice.
7. The social hierarchy found in many animals that live in groups.
8. The movement of entire populations of species between two areas, often at similar times of the year.
9. Kitti's hog-nosed bat (bumblebee bat). This species is found in south west Thailand. Its head and body length is just 2.9 to 3.3 cm. (just over an inch). It weighs only 1.7 to 2 grams (about 0.07 of an ounce).
10. Seals and walruses etc.
11. The form adopted by an insect between moults.
12. The liver.
13. The Galapogos fur seal, the female of which averages 1.2 metres (3 feet 11 inches) in length and weighs about 27 kg. (60 lbs.).
14. The blue whale and the fin whale both produce low frequency pulses that have been measured up to 188 decibels.
15. This modified wing is believed to provide information on stability in flight.

Nature Quiz

1. The largest colony of animals ever found was of black-tailed prairie dogs in 1901. How many animals were there and what was the estimated area covered by them?
2. Which fish species produces the largest egg?
3. Which is the fastest snake in the world?
4. What are chelonia?
5. What is the largest species of lizard in the world?
6. What species of wild bird is the most abundant in the world?
7. Why is the name of the horseshoe crab somewhat misleading?
8. What mammal blows bubbles to help satisfy its enormous appetite?
9. What is the act of ecdysis?
10. What is meant by the term 'amphibian'?
11. How long is the pregnancy of the elephant?
12. What is a pileus?
13. What shark learns to hunt its prey even before it is born?
14. What is plankton?
15. What is meant by a 'life cycle'?

ANSWERS

1. 400 million individuals which covered an area almost the size of the Republic of Ireland (61,400 square kms., that's 24,000 square miles).
2. The whale shark. The largest egg of this species was found in the Gulf of Mexico in 1953. It was 30.5 cm. by 14 cm. by 8.9 cm. in size. The live embryo inside measured 35 cm. long.
3. The black mamba, which is found in Africa. It can travel at between 16 and 19 km/h. (10 to 12 mph.) in short bursts.
4. Tortoises and turtles.
5. The komodo dragon. A male averages 2.25 metres in length (7 ft. 5 ins.). They have been reported to kill adult water buffaloes and human beings.
6. The red-billed quelea of Africa with an estimated adult breeding population of 1.5 billion.
7. This species is not really a crab at all but a distant relative of scorpions and spiders. The horseshoe crab is one of the oldest creatures on earth, little changed in 200 million years.
8. Humpback whales use rings of bubbles as a type of underwater net to trap fish and force them to the surface.
9. The shedding of the outer layer of skin in reptiles and the shedding of the exoskeleton in arthropods to allow for growth. A form of moulting.
10. Animals that can live both in water and on land.
11. Twenty-two months. The longest of all mammal pregnancies.
12. The cap of a mushroom or toadstool.
13. The sand tiger shark. During its year in the womb the dominant infant eats its brothers and sisters and when born is already one third of its mother's length.
14. Micro-organisms that float in surface waters of seas and lakes.
15. The various stages an organism passes through, from fertilised egg in one generation to fertilised egg in the next generation.

Nature Quiz

1. How heavy is the new-born calf of a blue whale?
2. What do the letters S.P.A. stand for?
3. What is a gall?
4. Why do moles make mole hills?
5. What is a gley?
6. What is the I.P.C.C.?
7. Why are compass termites so called?
8. Where did mankind get the idea for making paper?
9. Why is the tailorbird so called?
10. What is a troglobite?
11. What feather was considered sacred to the native American Indians?
12. How long can a tapeworm grow?
13. What bird requires at least half its own body weight of food each day?
14. What is a Portuguese man-of-war?
15. What colour are the flowers of bog rosemary?

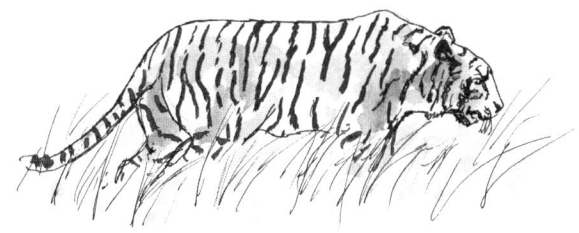

ANSWERS

1. A new-born calf can weigh up to three tonnes.
2. Special Protection Areas. These areas are preserved, under European Union directive, because they are particularly important for birds.
3. An abnormal swelling on a plant that is caused by a parasitic attack.
4. As air-breathing mammals they need to breathe. The mole hills are ventilation shafts allowing fresh air to circulate through their tunnels.
5. A waterlogged soil lacking in oxygen.
6. The Irish Peatland Conservation Council, 119 Capel Street, Dublin 1.
7. Because of their habit of building their earth mounds in such a way that the longer sides face east and west, which helps prevent overheating within the colony.
8. From the paper nests of certain wasp species. These insects chew wood and make a paper-type pulp to build their nests.
9. The long-tailed tailorbird makes its nest by sowing together large leaves on trees or shrubs. The female uses her beak as a needle and cobweb silk as her yarn.
10. A cave dweller. These creatures spend their entire lives in darkness and most species will have lost the use of sight.
11. The eagle's feather.
12. Some tapeworms have been known to grow to 15 metres long (50 feet).
13. A hummingbird. Its metabolic rate, with the possible exception of some shrews, is the highest of any known animal.
14. A jellyfish, actually a complex colony of individuals all developed from a single egg.
15. Pink or white.

Nature Quiz

1. What is the name of an otter's home?
2. Where would you normally find limpets?
3. What is the resting place of a hare called?
4. Where in the world do the longest periods of totality (complete darkness) occur during an eclipse?
5. How long does it take for the earth to orbit once round the sun. Is it one day, one month, or one year?
6. What fish do puffins predominantly feed on?
7. What are newly moulted crabs nicknamed?
8. Are wood ants found in Ireland?
9. What is a bittern?
10. What is an early purple?
11. What does the term 'roding' mean?
12. In relation to cetaceans, what does a 'rooster tail' describe?
13. What is a ruddy darter?
14. What are *Daphnia* species?
15. What disease is the tsetse fly a carrier of?

ANSWERS

1. A holt.
2. On the sea shore. They are shore-dwelling creatures found clinging to rocks.
3. A form.
4. The equator.
5. One year.
6. Sand eels.
7. Peelers.
8. Yes. They form colonies which are easily recognisable because of the mounds of dry plant material (leaves and/or pine needles).
9. A large brown wading bird closely related to the heron. Rare in Ireland, it is easily overlooked due to its habit of hiding in reedbeds.
10. A widespread and locally common orchid.
11. 'Roding' is the patrolling flight of the male woodcock as he flies just over the canopy of the woodland area of his territory. The bird whistles and grunts every few seconds and this is believed to attract available females.
12. The spray of water coming off the head of small cetaceans and is formed as the animal surfaces at high speed.
13. A dragonfly with reasonable distribution throughout Ireland.
14. Water fleas.
15. These blood-sucking flies carry trypanosome diseases such as sleeping sickness.

Nature Quiz

1. Apart from humans, what is the chief predator of sea mussels?
2. What do millipedes eat?
3. How many legs does a harvestman have?
4. What is a skipper?
5. What is an 'altruistic' relationship?
6. What colour is the flower of bladder campion?
7. Brambling, bullfinch, house martin, jay and wheatear all have a distinctive common feature. What is it?
8. Grayling is the common name of what two organisms?
9. Tawny owl, with its famous call 't-wit-t-woo', is the most common and familiar owl in Britain and Ireland. True or false?
10. Cape pigeon was the common name given by sailors to which seabird?
11. Potter wasps make separate vase-shaped nests from clay and mud for each of their offspring. True or false?
12. What colour is the bill of an adult sandwich tern?
13. Green tiger beetles are the sprint champions of the world. Do they occur in Ireland?
14. What is an eyas?
15. What is a small heath?

ANSWERS

1. The common starfish.
2. Millipedes are herbivores, eating living and dead plant material.
3. Eight. It is a spider.
4. One of the most primitive and moth-like of all butterflies. In Ireland our only species is the dingy skipper.
5. The act of one individual that increases the chances of survival of another, often offspring of a close relative, to the possible detriment of itself. A bee's sting, for example, benefits the colony but the bee dies after stinging.
6. White.
7. They all have a white rump.
8. Grayling refers to both a species of butterfly and a species of fish. The butterfly has a coastal distribution in Ireland whereas the fish, which is a freshwater animal, is common in Britain but not present in Ireland.
9. False. The facts are correct for Britain but this bird is not present in Ireland.
10. The cape petrel or pintardo.
11. True. These black and yellow solitary wasps make highly complex individual nests for each offspring. These nests are stocked with small caterpillars or other larvae – food for the grub.
12. Black with a yellow tip.
13. Yes.
14. A special term used by falconers for a nestling falcon or hawk.
15. A butterfly.

Amazing Facts

Pupfish Victory
The Devil's Hole pupfish is a small fish about 38 mm. (one and a half inches) long. It only lives in the Devil's Hole in the middle of the Nevada desert, USA. It feeds on minute creatures that collect on an algae-covered ledge. These fish made legal history in the USA when their existence was threatened by a scheme to pump water some distance away. It would have lowered the water table putting the feeding ledge at risk. The matter was brought to the United States Supreme Court. The Court ruled in favour of the pupfish, pumping was stopped, the pupfish was saved and legal history was made.

Crowded Space
40,000 different species can live in a single hectare of South American rainforest.

Ageing Turtles
Sea Turtles have roamed the oceans for at least 150 million years.

Seed Dispersal
The South American fruit bat is helping to combat deforestation. They regenerate rainforests by dropping seeds as they fly.

On The Wing
One of the most famous and spectacular feats of migration was completed by a manx shearwater that was removed from her breeding burrow and ringed on Skokholm, south Wales. As part of a navigational migration experiment the bird was taken to Boston, USA where she was released. Twelve and a half days later the bird was found back in her burrow on Skokholm having travelled 5,000 km. (3,125 miles).

Watery Planet
We take water for granted but did you know that 95% of the earth's water is actually chemically bound in rock. Of the remaining 5%, 97.3% is in our oceans, 2.1% is in the polar icecaps and glaciers and the remainder is fresh water. The fresh water is made up of atmospheric water vapour, ground water, soil water, and inland surface water.

Deer Quiz

1. What are male, female and young red deer called?
2. What are male, female and young fallow deer called?
3. What are male, female and young sika deer called?
4. Which is the largest species of deer: the red, fallow or sika?
5. What is a hummel?
6. What is the target?
7. Why are some fallow deer called menil?
8. What is a rut?
9. How many females could a red deer have in his harem?
10. What is a yeld?
11. St. Kevin is associated with the 'deer stone'. What county is it in?
12. In the Walt Disney cartoon film *Bambi*, what species was the deer?
13. A deer with a twelve-pointed antler is known as what?
14. Do deer lose their antlers every year?
15. The calls of the different species of male deer are known as what?

ANSWERS

1. The male is called a stag, female a hind and young a calf.
2. The male is called a buck, female a doe and young a fawn.
3. The male is a stag, female is a hind and young a calf.
4. The red deer (sika is the smallest of the three).
5. A male red deer (stag) without antlers.
6. The white or pale-coloured rear end of a deer beneath the tail. Also called the speculum or caudal disc.
7. Because they retain their spots throughout the winter.
8. The courting season for deer.
9. There can be as many as 50 females (hinds) in his harem.
10. A calfless red hind.
11. Glendalough, Co. Wicklow.
12. Red deer.
13. A royal stag.
14. Yes.
15. A fallow buck 'groans', whereas a red stag 'roars' or 'bells', a sika stag 'whistles', a roebuck 'bells' or 'barks' and a muntjac buck 'barks'.

Nature Quiz

1. How many feet does a common starfish have?
2. What is a faecal sac?
3. What is the other common name for a cleg?
4. Birdwatchers commonly refer to small unidentified birds, in slang, as LBJ's. What do these initials mean?
5. What is a coypu?
6. Where do glass frogs live?
7. One hare species is called the snowshoe hare. Why?
8. What bird has been dubbed the 'sentinel of the marsh'?
9. What is the relationship between the small waved umber and traveller's joy?
10. The 'rushing ceremony' is part of the courtship ritual of which bird?
11. What is the amplexus?
12. What is a brown mesite?
13. What is the '20Hz signal'?
14. What is another name for oarweed?
15. What colour are the flowers of black medick?

ANSWERS

1. Five.
2. The tough mucous sac containing the nestling's faeces. These sacs are produced to assist nest sanitation allowing the parent bird to remove the dropping simply and cleanly.
3. The horse-fly.
4. Little brown job.
5. A large aquatic rodent, a native of South America. There are various wild populations established around Europe as a result of escapes from fur farms. A long established population persists around the Norfolk broads in eastern England.
6. There are at least 65 species of these tree-dwelling amphibians which live in the cloud forests of the Andes. They are called glass frogs because, although generally green, the skin on their undersides is so transparent that you can see their internal organs.
7. Although brown during the summer months, the snowshoe hares of Alaskan and Canadian forests turn completely white in winter (except for black tips on their ears). They also grow thick pads of fur on the soles of their feet (their 'snowshoes'), which insulate and help with movement on snow.
8. The redshank, because of its harsh, three-note alarm call.
9. Traveller's joy is the food plant of the small waved umber moth's caterpillar.
10. The western grebe of North America (a close relation of our own great crested grebe).
11. The mating posture adopted by frogs and toads. With common frogs and toads in these islands it is usually completed within a day, but prolonged amplexus lasting several months is known in frogs in the high, cold regions of the Andes.
12. A bird species endemic to the evergreen forests of Madagascar and very little is actually known about this bird or its two closely-related species.
13. For years the source of this sea sound was a mystery until it was discovered that it is produced by the fin whale, whose song, used for communication, is audible at a distance of 800 km. (500 miles).
14. The other name for this common seaweed is kelp.
15. Yellow. The ripe pods are black.

Nature Quiz

1. What creature has over 2,000 eyes?
2. What four animals are the main characters in Kenneth Grahame's book *The Wind in the Willows*?
3. What is a turlough?
4. Where would you find a 'sticky mouse-ear'?
5. Woodpeckers are found in Ireland. True or false?
6. What does the word 'hypogean' mean?
7. What strange way does the Texas horned lizard defend itself against attack?
8. What is a hyperparasite?
9. What is a garden pebble?
10. What type of creature is believed to have the largest eyes in the world?
11. What is a baby blue shark called?
12. In relation to body-length the tail length of a red kite is proportionately longer than the tail length of a long-tailed tit. True or false?
13. What is the expected life span of a termite queen?
14. What is infrasound?
15. Killer whales live in family groups known as pods. What are closely-related pods called?

ANSWERS

1   A dragonfly. Its two large and bulging compound eyes cover almost its entire head. Each eye is made up of more than 1,000 tiny eyes; six-sided facets, each with its own lens and retina.
2   Mole, Rat, Toad and Badger.
3   A specialised habitat formed by a shallow limestone depression that floods when the water table rises after autumn and winter rainfall.
4   This common and widespread plant is found on dry bare ground. It flowers from April to October.
5   True. Although not resident in this country, some species have been recorded here. e.g. The great spotted woodpecker, a rare migrant from Scandinavia.
6   Growing or occurring underground.
7   When threatened this lizard bursts a number of small blood vessels in its eye membranes, squirting jets of blood at its attacker.
8   A parasitic insect that lives in or on a host that is itself parasitic on another species of insect.
9   A small moth of the Pyralidae family.
10 The giant squid, when fully grown, measures up to 18 metres (60 feet) and has eyes the size of footballs.
11 A pup.
12 True.
13 Secure in the royal chamber of a busy termite colony, the 15 cm. (6 inch) long egg-laying queen should live for fifty years.
14 Infrasound is below the range of human beings, though it is often 'felt' as vibrations. Scientists have speculated that migrating birds may be able to use infrasound 'landmarks' as guides on their route.
15 They are known as clans and even develop their own dialects.

Nature Quiz

1. What is bioremediation?
2. What is pollination?
3. What are callows?
4. Which animal is celebrated in William Blake's poem from his *Songs of Innocence*?
5. Rare alpine, arctic and Mediterranean plants are found in the Burren. True or false?
6. What is the technical word for a body of standing water that becomes 'nutrient enriched'?
7. When a habitat is described as 'diverse', what does it mean?
8. Coots, finfoots, grebes, and phalaropes, share one particular adapted characteristic. What is it?
9. Occasional 'irruption' years of waxwings, from their breeding grounds in Scandinavia, allow us the chance of seeing these delightful, rare winter visitors to Ireland. The failure of what berry crop is linked to these irruption years?
10. The little egret is currently expanding its breeding population in Ireland. What colour are its legs and feet?
11. Scots pine was originally a native tree, but died out. It was re-introduced in approximately what year?
12. Where would you look for blewits?
13. The term 'fast ice' refers to what?
14. The lower leaves of a holly tree have prickly spines around the edge, whereas the leaves near the crown are plain oval and are spineless, why?
15. What is the common name for the baldpate?

ANSWERS

1. This recently developed technique is already proving effective in dealing with oil spillages. The method uses one of the world's oldest pollution controllers – oil-eating marine bacteria.
2. This is when the pollen from a male flower reaches the female ovule and causes the fertilisation of a plant.
3. Callows are the natural meadows in the floodplains of large rivers. The Shannon callows is one of the few remaining areas where you can still hear the corncrake.
4. The Tyger (tiger).
5. True. Their presence in the Burren is still a puzzle to botanists.
6. Eutrophication.
7. It is used as a descriptive measure of the variety of species present in a habitat.
8. They all have lobed (lobate) feet. This is where each toe is separately fringed by lobes (as distinct from webs connecting the toes), to assist the bird in swimming.
9. The rowan tree (mountain ash). The successful berry crop years are followed by a poor crop year and, apparently, waxwings anticipate the pending shortage of food and move south and west.
10. This small white heron species has black legs and bright yellow feet.
11. 1652.
12. Blewits, a toadstool, is found mainly in pastures and occasionally near hedges.
13. Sea ice attached to the shore or stretching between grounded icebergs.
14. Another of nature's defence mechanisms. The lower leaves are at risk of being grazed by animals, whereas near the top of the tree there is little risk so the need for that sort of protection is not required.
15. The American wigeon. A rare, but almost annual, visitor to Ireland.

Nature Quiz

1. What is leucism?
2. In an aquatic habitat, where do the surface dwellers live?
3. What is a proboscis?
4. The word 'invertebrate' means exactly what?
5. What are the most common insect warning colours?
6. Which Government Department is responsible for wildlife in Ireland?
7. What is the name given to the shed larval skin of a dragonfly?
8. What does the word 'murine' mean?
9. Where does navelwort grow?
10. What is a redstart?
11. What colour are the fruits of cloudberry?
12. What causes false rings in timber?
13. What is a polypody?
14. Both the blackbird and song thrush make deep cup-shaped nests that are carefully hidden in a variety of habitats. One species lines the nest with wood pulp or mud whereas the other species lines it with grasses. Which species has the nest lined in wood pulp or mud?
15. What is the horny (calcareous) plate used by a snail to close the mouth when it withdraws into its shell?

ANSWERS

1. This rare phenomenon, in a species, is the partial loss of pigment which affects the colours by reducing them in intensity. Complete loss of pigmentation is albinism, where the pigment is absent throughout the complete body.
2. Actually on the surface film of the water.
3. The elongated mouth parts of some insects which is used to obtain food. e.g. the humming-bird hawk-moth can obtain nectar from deep within a plant, through the proboscis, whilst hovering in front of it.
4. An animal without a backbone.
5. Yellow and black, orange and black and red and black.
6. The Department of Arts, Heritage, Gaeltacht and the Islands, at Mespil Road, Dublin 4, has the section of the National Parks and Wildlife as part of Dúchas The Heritage Service.
7. Exuvia.
8. It means of, related to, or similar to, a mouse. Can also mean mouse-coloured.
9. It grows on walls and stony banks, often in partial shade.
10. It is a rare summer migrant bird. The name literally means red tail; start being the old English word for tail.
11. Orange when ripe.
12. False rings in timber look like annual rings, but are not complete. They may be caused by severe frost, defoliation or some other damage to the tree.
13. Polypodies are ferns often found on the trunks of mossy trees, rocks and walls. They prefer drier areas than other ferns.
14. The song thrush.
15. The operculum.

Nature Quiz

1. The gargany is our only summer-visiting duck. Where does it winter?
2. What are antennae?
3. What is a katabatic wind?
4. Blubber is the layer of insulating fat just under the skin of cetaceans. In some species it can be as thick as 50 cm. (20 inches). True or false?
5. Grasshoppers and crickets are insects that have an incomplete metamorphosis. Which stage is missing?
6. What is a sea mouse?
7. Which introduced redwood tree in Ireland is renowned for its soft bark?
8. What colour are the flowers of sea bindweed?
9. Painted sweetlips are what?
10. Dust mites are a common everyday feature of the normal household. How many would the average double bed be home to?
11. Where do shelduck go to moult in the autumn?
12. Sparrow species are sexually dimorphic except for one species. In which species do the sexes look the same?
13. What is the jet stream?
14. What is the offspring of a male lion and a female tiger called?
15. What is the offspring of a male tiger and a female lion called?

ANSWERS

1. In Africa. Some in the Mediterranean basin but the large majority in West Africa.
2. The movable accessories, on the heads of some animals, that are sensitive to taste and touch.
3. A wind experienced in the Antarctic, driven by gravity and caused by colder, heavier air rushing down from the polar plateau.
4. True.
5. The pupal stage. They develop from an egg through a variable number of nymphal stages to the adult.
6. An oval-shaped worm, covered with brownish-grey hairs, found on the lower sea shore. The longer golden-brown and greenish bristles, along its sides, can give off an iridescent sheen.
7. The wellingtonia, also known as the giant sequoia. Its soft bark can be punched with impunity. It is well worth looking for the indentations made by roosting treecreepers.
8. Pink, with white stripes to the edges from the centre.
9. A fish found on the Great Barrier Reef, Australia.
10. Believe it or not about 2 million!
11. In the Heligoland Bight area of the North Sea and to Bridgewater Bay in Somerset.
12. The tree sparrow.
13. A narrow band of high-speed, westerly winds found at heights of between 9,000 to 15,000 metres (30,000 to 50,000 feet). Maximum velocities are normally 160 km/h. to 240 km/h. (100 to 150 mph.) but may increase to 480 km/h. (300 mph.) in winter.
14. A liger.
15. A tigon.

Nature Quiz

1. What is the abdomen of an insect?
2. What is a mammary gland?
3. How did the pilotfish get its name?
4. Caoutchouc is collected from a variety of tropical trees. What is it?
5. *Ursa Major* and *Ursa Minor* are two constellations close to the pole star in the night sky. Which animal do they symbolise?
6. How does the handkerchief tree get its name?
7. What is salinity?
8. What is the tallest cactus in the world?
9. What is the weight of the average ostrich egg?
10. What is a billabong?
11. What introduced plant species is causing serious damage to our own native woodlands?
12. The killer whale is the largest dolphin. True or false?
13. What does the term 'endemic' mean?
14. What does coniferous mean?
15. What bird is associated with the Tower of London?

ANSWERS

1. The rear section of an insect's body.
2. In female mammals it is a gland responsible for the production and release of milk to feed their young. They vary in number from two to twenty.
3. From the behaviour of young pilotfish that swim alongside ships or large fish, such as sharks.
4. Raw rubber – obtained from the milky latex exuded by the trees as a response to injury.
5. The bear. The names translate to Great Bear (also known as the Plough) and Little Bear.
6. This deciduous tree, native to China, has tiny flower clusters protected by two large, white, pendulous bracts, which give the tree its name.
7. The saltiness of sea water. The amount is usually expressed as parts per thousand ($^0/_{00}$) grams of solute per kilogram of seawater. The average salinity of ocean water is about $35^0/_{00}$.
8. The saguaro. This slow-growing cactus can reach 21 metres (70 feet) and is found only in Arizona, Southern California and the Sonoran Desert in Mexico.
9. It weighs about 1.5 kg. (3 lbs.).
10. The Australian term for a stagnant pool of water.
11. Rhododendron, which is growing at an alarming rate in the wild and displacing some of our native woodlands. e.g. In Killarney and Glenveigh National Parks.
12. True.
13. A term applied to a species in relation to a stated area that it is restricted to.
14. Cone-bearing.
15. The raven. Superstition has it that should the ravens ever abandon the Tower of London, disaster will follow.

Nature Quiz

1. At what time of the year does the solstice happen?
2. Which is the largest carnivore of Ireland and Britain?
3. We take sunlight for granted. How long does it take sunlight to travel from the sun to the earth?
4. What is the 'Chinese gooseberry' better known as?
5. What is currently considered as the first bird known to have existed?
6. Which species of bird is considered as laying the most 'colour and pattern-varied egg' of all birds?
7. What is unusual about a basenji dog?
8. What is a legume?
9. What cat has no tail?
10. How did earwigs get their name?
11. The southernmost penguin rookery in the world is situated at Cape Royds, Ross Sea. Which small penguin breeds there?
12. What is a sea-hare?
13. What colour is corn marigold?
14. Sharks are a large group of predatory fish of varying sizes. What length is the smallest of these creatures?
15. What bird is nicknamed 'the laughing jackass'?

ANSWERS

1. The shortest day and the longest day of the year. It is when the sun is furthest from the celestial equator. That is, when it is at the most northern point in winter and southern point in summer.
2. The grey seal.
3. Eight minutes.
4. The kiwi fruit.
5. *Archaeopteryx*, which translated means ancient wing. Considered to have lived about 150 million years ago, it could fly, was warm-blooded and was covered in perfect feathers.
6. The guillemot. Colour varies from white through, creams, yellows, browns, blue-greens and blues. They also have a very wide variety of patterns and squiggles.
7. It cannot bark.
8. A pod. It is the fruit of the members of the *Leguminosae* family, which include the peas, beans, gorse and laburnums.
9. The Manx cat.
10. The origin of the name earwig is in the likeness of its pincers to the old-fashioned instrument used to pierce ears for earrings.
11. The Adélie penguin.
12. A slow-moving mollusc found in rock pools on very low tides. The sea-hare has the ability to eject a purple dye if alarmed.
13. As its name implies, yellow.
14. The smallest shark, *Squaliolus laticaudus*, which is found off the Philippines, is just 15 cm. (6 inches) long.
15. Kookaburra. The blue-winged kookaburra also has an alternative nickname of howling jackass.

Junior Quiz

1. What are a bird's feathers called? _The plumage_
2. What is a dromedary? _Camel with 1 hump_
3. How many hours does a koala sleep in a day? Is it 2 or 12 or 18 hours? _18 hours a day_
4. What is the correct name for the American buffalo? _bison_
5. Which is the largest member of the crow family? _falcon_ x
6. *Maiasaura* was a dinosaur. What does its name mean? _good mum_
7. What is a roadrunner? _A bird in country in america_
8. What is an oasis? _A lake of water in a dessert but it's not real_
9. The zebra is a member of the horse family. True or false? _True_ ✓
10. What type of water is home to the brine shrimp? _Salt water_
11. How can you tell the age of a cut-down tree? _how much rings_
12. What tree would you get conkers from? _horse chestnut_ ✓
13. What owl in Ireland and Britain is known as a screech owl? _Barn owl_
14. Why do some seaweeds have air sacs in their leaves? _to help float_
15. What colour is the flower of the ox-eye daisy? _yellow and white_ ✓

6/15

ANSWERS

1. The plumage.
2. An Arabian camel with one hump (the bactrian camel has two humps).
3. 18 hours in a day.
4. The bison.
5. The raven.
6. Good mother.
7. A bird found in open country in North America. This bird is related to the cuckoo.
8. A place in desert regions where water is present.
9. True. It is the only horse that has stripes.
10. Salt water.
11. By counting the rings from the centre of the trunk outwards.
12. The horse chestnut.
13. A common name for the barn owl, because of its piercing scream.
14. These air sacs (or bladders) help some seaweeds to float in the water.
15. The flower has white petals with a yellow centre.

Junior Quiz

1. What tree produces acorns? __Oak__
2. The wren is the smallest bird in Ireland. True or false? __True__
3. What is a boar? __A male pig__
4. What is a biennial plant? __A plant with two cycles__
5. What is meant by 'camouflage'? __a animal that blend into the background__
6. In a wildlife sense, what does the term native mean? __usual in a area__
7. The robin is a member of the thrush family. True or false? __True__
8. What is a cub? __a baby lion or tiger or any wild cat__
9. What is a male rabbit called? __buck__
10. What is a male duck called? __drake__
11. What is an embryo? __young plant__
12. What is a finch? __small seed eating bird__
13. How many species of squirrel can you find in Ireland? __2__
14. What are the five principal colours of the rainbow? __red, blue, violet, yellow, green__
15. How does the dandelion spread its seeds? __The seed are attached to little parachuters then they get blown away.__

ANSWERS

1. The oak tree.
2. False. Although the wren is widely believed to be the smallest, it is actually the goldcrest and its close, but rare, relative the firecrest that are Ireland's smallest birds.
3. A wild pig. Widely distributed in forests of Southern and Central Europe, and used to be common in Britain but is now extinct.
4. A plant that has a two year cycle. The first year it stores the food and in the second year it uses up the food producing seeds. The plant then dies.
5. A means by which various animals avoid detection, often by having a coloration that matches their background.
6. This is a species that occurs naturally in an area and therefore one that has not been introduced either accidentally or intentionally.
7. True.
8. They are the young of certain mammals such as the fox and badger.
9. A male is a buck.
10. A drake.
11. A young plant or animal still in its early stages of development.
12. Finches are small to medium-sized seed-eating birds. Greenfinch, chaffinch, goldfinch, bullfinch, redpoll and siskin are all members of this large family.
13. Two. The red squirrel and the grey squirrel.
14. Red, yellow, green, blue and violet. There are mixed hues where the colours overlap.
15. The seeds are each attached to their own little 'parachute'. When conditions are right, the wind blows the seeds away.

Photo
Quiz

1 In folklore, what is the name given to the seal people?

2 Which is the largest kangaroo in Australia?

Answers: 1 Seal fairies are known as 'selkies' around Scottish Islands and are called 'roane' in parts of Ireland. 2 The red kangaroo.

Photo Quiz

1. A relative of the goldcrest, what is this bird called?

2. Antarctic resident, can you name it?

3. A graceful bird of lakes and rivers, what is it?

Answers: 1 Firecrest. 2 Adélie penguin. 3 Mute swan.

Photo Quiz

1. Found in Britain and Europe but absent from Ireland, what is this owl?

2. A clue to this lovely white flower is 'snow falling'?

3. Name this attractive moth?

Answers:
1. Little owl.
2. Snowflake.
3. Emperor moth.

Photo Quiz

1. What is an important part of a badger's diet?

2. Name this beautiful orchid?

3. The butcher bird. What is its real name?

4. The bill is the clue to the name of this bird?

Answers:
1 Earthworms
2 Bee orchid.
3 The shrike. This bird is the great grey shrike.
4 Crossbill.

Photo Quiz

1 Introduced into Ireland. Can you name this deer species?

2 What is a young hare called?

Answers:
1 The sika deer.
2 Leveret.

Photo Quiz

1 Lord of the mountain. Can you name this bird of prey?

2 Common in Britain, making a comeback in Ireland. What is this large raptor?

Answers: 1 Peregrine falcon. 2 Common buzzard.

Photo Quiz

1 Garden friend, not foe. What is this spider?

2 What is this moth?

3 Name this petrel found in the Antarctic?

4 Our only reptile. Can you name it?

Answers:
1 Garden (or cross) spider.
2 Wood tiger.
3 Snow petrel.
4 Viviparous lizard (also known as common lizard).

Photo Quiz

1 Jewel of the river. What is this bird?

2 Sometimes called 'the swallows of the sea'. What are these elegant birds?

3 Name this cliff-nesting seagull chick?

Answers:
1 Kingfisher.
2 Terns. *Shown are two roseate terns.*
3 Kittiwake.

Junior Quiz

1. The greenfinch is predominantly two colours. What are they? Green and yellow X
2. What causes tides? heavy rain and wind
3. What is a female rabbit called? doe X
4. What colour is the female blackbird? brown X
5. How many legs does a spider have? 8
6. What flower, when held under the chin, is said to show whether that person likes butter or not? buttercup
7. What bird lays its eggs in other birds' nests? cuckoo
8. Ice is heavier than water. True or false? False X
9. Mallard, eider, wigeon and teal are all types of what? Duck X
10. What is the science of star-gazing called? astronomy
11. Should you feed wildbirds in the summer as well as the winter? Yes
12. Elm, hornbeam, alder and sycamore are all types of what? tree
13. Do moths only fly at night? No
14. Can hazel nuts be found in the wild in Ireland? Yes
15. What common bird is associated with Christmas? Robin

10/15

ANSWERS

1. Green and yellow.
2. The moon's gravitational pull.
3. A female rabbit is called a doe.
4. Brown, sometimes with quite a pale patch under the chin.
5. Eight.
6. The buttercup.
7. The cuckoo.
8. False. Ice floats in water.
9. Duck.
10. Astronomy.
11. Yes. It helps the adult birds keep in good condition while they bring up their families.
12. Deciduous trees.
13. No. There are a number of species that fly during the day.
14. Yes. The hazel is common in Ireland.
15. The robin.

Junior Quiz

1. How many legs does a butterfly have? __6__ ✓
2. In the book *Alice in Wonderland*, what did the caterpillar tell Alice to eat to grow taller or shorter? __mushroom__ ✓
3. What organisation, in Ireland, looks after distressed or injured seals? __Irish seal sanctuary__ ✗
4. What bird is associated with St. Stephen's day? __Wren__ ✗
5. What is the fruit of the bramble called? __blackberry__ ✗
6. Some people have a fear that a bat could caught in their hair. Does this ever happen? __sometimes__ ✗
7. Where does a skylark nest? __ground__ ✗
8. How many wings do true flies have? __2__ ✓
9. If a bird is 'preening', what is it doing? __cleaning feathers__
10. Are there any wild snake species in Ireland? __no__ ✓
11. A blackcap is a warbler. True or false? __True__ ✗
12. What is a young frog called? __tadpole__ ✓
13. What birds are said to have a passion for collecting shiny objects? __magpie__
14. What sound can be heard just before sunrise in spring and early summer? __birds singing__ ✓
15. In Irish mythology, what fish is said to have given Fionn the gift of knowledge? __salmon__ ✓

8/15

ANSWERS

1 Six.

2 A mushroom.

3 The Irish Seal Sanctuary.

4 The wren.

5 The blackberry.

6 No. Bats are far too skilled to fly into large objects.

7 On the ground, in meadows and open areas.

8 Two.

9 Cleaning its feathers. This exercise is vital to the bird's existence.

10 No.

11 True.

12 A tadpole.

13 The jackdaw and magpie.

14 The dawn chorus. Birds sing to proclaim their breeding territories.

15 A salmon.

Junior Quiz

1. Speckled wood, meadow brown, and large heath are all types of what? butterflies X
2. What bird species plays the central role in *On Silent Wings* by Don Conroy? Owl ✓
3. What is a chrysalis? stage of insects Life X
4. What plant is said to relieve the pain from the sting of a nettle? Dock leave ✓
5. What was the name of the dolphin that starred in the popular American TV series of the same name? Jungle X
6. Perch, bream and roach are all types of what? fish X
7. What was the name of the kangaroo that starred in the popular Australian TV series of the same name? Joey X
8. What is a male mute swan called? cob ✓
9. What rodent made Walt Disney famous? mouse ✓
10. The adult small tortoiseshell butterfly hibernates in the winter. True or false? false X
11. Do strawberries grow wild in Ireland? yes ✓
12. What are an elephant's tusks made of? Ivy ✓
13. In Steven Spielberg's movie *Jurassic Park*, which dinosaur learned how to open doors? T. Rex X
14. What is an animal's spoor? footprints X
15. Hares and rabbits often share the same burrow. True or false? True X

5/15

ANSWERS

1. Butterflies.
2. The barn owl.
3. The stage in an insect's life cycle between the caterpillar (larva) and the adult.
4. The broad-leaved dock.
5. Flipper.
6. Fish.
7. Skippy the bush kangaroo.
8. A cob.
9. A mouse known as Mickey.
10. True.
11. Yes. The wild strawberry is common in Ireland along hedges and shady banks. The fruit are much smaller than commercially grown strawberries.
12. Ivory.
13. The velociraptor.
14. The animal's footprints.
15. False. Rabbits live in burrows but hares never go underground.

Junior Quiz

1. What type of shark featured in the film *Jaws*? white shark ✓
2. Do all owls hunt at night? Yes ✗
3. What is pollution? Litter in rivers ✓
4. What is a bird pellet? prey birds, owls ✗
5. What dogs are used to pull sledges? Huskeys ✓
6. Name the two types of rhinoceros found in Africa? Black + White ✓
7. Nightingales sing only at night. True or false? False ✓
8. What is the skin of a tree called? bark ✓
9. What is the biggest insect in the world? giant stick insect
10. The Niagara Falls border what two countries? Canada + U.S.A
11. What seal-like creature has two prominent tusks? sea lion ✗
12. What is a common name for thrift? pink ✗
13. Where would you find the Giant's Causeway? Donegal ✗
14. What is a female mute swan called? pen ✗
15. What colour is the new-born pup of an Atlantic grey seal? white ✓

7/15

ANSWERS

1. Great white shark.
2. No. Some, like the short-eared owl, can be seen hunting in daylight hours.
3. Anything that dirties or destroys the environment. e.g. Exhaust fumes, litter, chemical waste, etc.
4. Certain birds, such as birds of prey and owls, cannot digest fur and bones and cough up these remains in what is called a pellet.
5. Husky dogs.
6. The black rhinoceros and the white rhinoceros.
7. False.
8. The bark.
9. The giant stick insect. It is 38 cm. (15 inches) in length.
10. Canada and the USA.
11. The walrus.
12. Sea pink. A small, pink-flowered seaside plant.
13. County Antrim in Northern Ireland.
14. A pen.
15. Creamy-white.

Junior Quiz

1. What happens when a snake moults? sheeds skin ✗
2. What is a pine marten? small mamel ✗
3. A hyena can munch its way through an entire wildebeest carcass – hide, horns, hooves and bones. True or false? true ✓
4. What are springboks, impalas and gazelles doing when they are 'pronking'? leaping in air ✗
5. What is a young mute swan called? cygnet ✓
6. What is a baby rabbit called? kitten ✓
7. What is the name of the rainy season in tropical countries? monsoon ✓
8. Are there any poisonous snakes in Britain? no ✗
9. What is a baby duck called? duckling ✓
10. Which comes first, thunder or lightning? thunder ✗
11. What does a house martin make its nest out of? mud ✓
12. Do wild roses have thorns? Yes ✓
13. Snow is made up of crystals of ice. Are these crystals all identical? no ✓
14. What colour is the shell of the rough periwinkle. Is it black, grey, brown, red, green, yellow or orange? brown ✗
15. The lesser weever fish lies in sand in shallow waters with only its eyes showing. Why is it a good idea to avoid it? It can sting

7/15

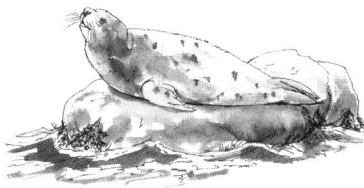

ANSWERS

1   It sheds its skin.

2   A small mammal related to the stoat and weasel.

3   True. It has exceptionally powerful jaws.

4   Leaping straight up in the air. They do this when they are being pursued by a predator.

5   A cygnet.

6   A baby rabbit is called a kitten.

7   The monsoon season.

8   Yes. The adder.

9   Duckling.

10 Lightning. Thunder is the sound of the lightning and as sound is slower than light, it is the lightning that is seen first. You can estimate the distance of the lightning by counting the gap between the flash and the thunder. A five second delay equals 1.6 km. (1 mile).

11 Mud.

12 Yes.

13 No. Every snowflake is made up of crystals and each crystal is unique.

14 It can be any one of these colours.

15 If you stand on this fish the spines can inflict a painful sting.

Junior Quiz

1. What bird is called 'the clown of the sea'? clownfish ✗
2. Pearls are found in what animals? oyster ✓
3. Where do herons build their nests? trees ✗
4. Can earwigs fly? some can
5. How good is an earthworm's vision? not very good ✗
6. What was a *Triceratops*? type of dinosaur
7. Why did miners bring canaries into coal mines? birds sensitivity
8. In Disney's animated movie *The Lion King* two of the central characters were called Timon and Poomba. What type of animals were they? meerkat + hog ✓
9. What is a baby elephant called? calf
10. All holly trees have red berries. True or false? True ✗
11. What causes hay-fever? flowers ✗
12. How many wings does a dragonfly have? 4 ✓
13. Herring, common and black-headed are all types of what? birds ✓
14. How does a python kill its prey? chokes them
15. What is a warren? maze of burrows

5/15.

ANSWERS

1. The puffin, because of its brightly-coloured bill and black and white plumage.
2. Oysters and mussels.
3. Normally in the tops of tall trees.
4. Some species can but others have no wings and therefore can't.
5. Dreadful – they have no eyes at all.
6. A three-horned dinosaur.
7. The bird's sensitivity to toxic gasses gave the miners an early warning of danger.
8. A meerkat and a warthog.
9. A calf.
10. False. Female holly trees bear berries whereas male trees do not.
11. The microscopic pollen of grasses floating in the air.
12. Four.
13. Gull species.
14. It winds its body around its prey and crushes them to death.
15. A maze of burrows inhabited by a rabbit colony.

Amazing Facts

Rainfall Cycle
The average turnover of atmospheric water is just 11.4 days which means that all water vapour in the atmosphere falls as rain (precipitation) and is re-evaporated more than 32 times every year.

Speeding Planet
The earth is travelling around the sun at an average speed of 30 km. (18.75 miles) per second.

Everything Was Nothing?
The total mass-energy of the Universe was apparently formed out of nothing, something which cannot yet be explained by anyone on this planet.

Free Meal
So powerful is a parent bird's instinct to feed the gape of a chick that a North American red cardinal finch once dutifully fed a goldfish in a pond. The fish had learned to stick its mouth above water to take food from its owners. On seeing the bird's shadow, it duly gaped. The finch instinctively shoved a beak-full of insects into the fish's mouth.

Time Tells
To a scientist, one second is the time taken for an atom of caesium-133 to oscillate 9,192,631,770 times.

Confused Poles
30,000 years ago Magnetic North was at the South Pole. This is because Magnetic North is continually moving and, since the earth was formed, magnetic fields have been reversed on a number of occasions.

Bat Quiz

1. How many species of bat occur in Ireland?
2. What is the smallest bat in Ireland?
3. What do bats eat?
4. Which bat species population in Ireland is of major international importance?
5. Are bats blind?
6. How do bats 'find their way' at night?
7. How many young will an adult bat produce in a year?
8. Do bats make nests?
9. Do bats hibernate?
10. What bat species is associated with Dracula?
11. How many bat species occur in Europe?
12. Is the bat the only true flying mammal?
13. Which bat species is called the flying fox?
14. How many midges is a pipistrelle bat capable of eating in a single night. Is it 100, 1,000, 3,000 or 5,000?
15. Are bats protected in Ireland?

ANSWERS

1. Seven. Pipistrelle, Leisler's, long-eared, Natterer's, Daubenton's, whiskered and lesser horse-shoe. (The pipistrelle bat has recently divided into two separate species).

2. The pipistrelle (it can fit into a matchbox).

3. All bats in Ireland and Britain eat insects (midges, crane flies, mosquitoes, earwigs, spiders, moths etc.).

4. The lesser-horseshoe bat.

5. No.

6. Bats produce a high pitched sound known as echolocation. The sound bounces off both stationary and moving objects and gives the bat a sound picture of its surroundings.

7. One. Young are usually born in June and July in the British Isles.

8. No. Female nurseries are usually in buildings. The young cling to the mother.

9. Yes.

10. The vampire bat.

11. Thirty-three species of which thirty-one are indigenous and two are irregular vagrants.

12. Yes.

13. The fruit bat.

14. 3,000.

15. Yes.

Wildlife Quiz

1. What is the equinox?
2. What is the national flower of Austria?
3. How many humps has a new-born camel?
4. What is listed in Red Data Books?
5. How can an eagle see sideways without turning its head?
6. What is the difference between common alder and grey alder?
7. What is a jaeger?
8. What is aetiology?
9. The brown hairstreak is a scarce butterfly found particularly in the Burren. Eggs are laid by the female in which hedgerow shrub/tree?
10. How does a baby snake get out of its egg?
11. Cuckoos are known for the fact that they lay their eggs in other birds nests. A female cuckoo can only lay an egg in the nest of the species that she herself was brought up by. True or false?
12. Yeasts are fungi. True or false?
13. What colour is the flower of the nettle-leaved bellflower?
14. What does 'precipitation' mean?
15. What is a blushing bracket?

ANSWERS

1 The moment when the sun crosses the equator and makes day and night equal throughout the world. The vernal equinox occurs on 21st March and the autumnal equinox occurs on 23rd September.
2 Eidelweiss.
3 None.
4 Species of 'concern' and 'endangered' species.
5 They have two focal points in each eye.
6 The common alder has almost black bark and rounded leaves, whereas the grey alder has grey bark and toothed triangular leaves.
7 It is the alternative name (and used mainly in America) for skuas of the *Stercorarius* group (the Arctic, pomarine and long-tailed).
8 The science of causation with particular reference to causes of disease.
9 The blackthorn.
10 Young snakes have an egg tooth which projects from the upper lip, and this is used to slice through the hard egg case. It is discarded as soon as the snake has emerged from the egg.
11 True. Although cuckoos can produce a range of varied egg patterns, each individual only produces one pattern. It is believed that this is inherited from her own mother.
12 True.
13 Violet-blue.
14 It means all the forms in which water falls to the ground. It includes rain, drizzle, sleet, snow and hail.
15 A common bracket fungus found on the dead branches of willow and birch.

Wildlife Quiz

1. 'Jugging' is a word that is given to the sleeping place of what bird?
2. What does 'palmate' mean?
3. A shark's eyes are ten times more sensitive to dim light than a human's. True or false?
4. How far can a red kangaroo travel in a single bound?
5. What is an amethyst deceiver?
6. What is a grey chi?
7. Where would you find rose-root?
8. Where, on a bird, would you find the cere?
9. What colour are the flowers of greater bladderwort?
10. A peacock can have more than 200 tail feathers. True or false?
11. When using echolocation, how many clicks can a dolphin produce in a second?
12. What is the main diet of an osprey?
13. The smallest marsupials are the insectivorous dasyurids or marsupial 'mice'. What is another common local name for these small animals?
14. What does the word 'pilose' mean?
15. In relation to volcanoes, what is tephra?

ANSWERS

1. The partridge. It is also used in relation to their call.
2. Finger-like.
3. True. They also have the ability to switch off their dark adaptation capabilities and function normally in brighter light.
4. 12.8 metres (42 feet). They can travel at 64 km/h. (40 mph.) on their hind legs.
5. An attractive, lilac or purple, fungus of deciduous woodland leaf litter.
6. A moth found in grassy uplands and on moorland.
7. A plant of mountain ledges and sea cliffs, its flowers appear in May and June and are rounded yellow clusters.
8. At the top of the bill. It is the skin at the base of the upper bill, particularly noticeable in birds of prey.
9. Yellow.
10. True. The more eyespots in his tail the better his chances of winning a mate.
11. Up to 700. Each of these are used in the analysis of the objects, prey and other dolphins around them. To our ear the different clicks become fused together as a continuous sound when just 20 to 30 clicks a second are produced.
12. Fish. Ospreys have particularly long legs and talons adapted to hold onto slippery fish that are caught by the bird plunging into water to 'grab' their prey.
13. Dunnarts. They weigh as little as 2 grams and can be just 9.5 cm. (3.75 inches) long.
14. It means covered with fine hair or down.
15. It is all the particles and fragments ejected from a volcano.

Wildlife Quiz

1. When a species is said to be indigenous, what does it mean?
2. 'Halcyon' is the poetic word for what bird?
3. Where would you find a common earth-star?
4. There are no marsupials living in the wild in England. True or false?
5. Toads have no teeth whereas frogs have very small teeth in the upper jaw only. True or false?
6. What is a dik-dik?
7. Which common plant, often found in lawns, can be used in salad, as a coffee substitute, and to make wine?
8. What is earthshine?
9. What is the smallest penguin in the world?
10. What is the boiling point of water on the Fahrenheit scale?
11. When is an animal species categorised as extinct?
12. What is falconry?
13. What is *El Niño*?
14. What is frankincense?
15. Where would you eat a fugu?

ANSWERS

1. That it occurs naturally and has not been introduced.
2. Kingfisher.
3. In woodlands. It is a fungus with fruit that looks not unlike an onion. The outer layer splits and the segments peel back, eventually lifting the main fruit off the ground.
4. False. There are two feral populations of red-necked wallaby living in England. One population in Sussex and one population living in the Peak District.
5. True.
6. A dwarf antelope, native to Africa reaching no more than 400 mm. (16 inches) in height.
7. The dandelion. The leaves of this perennial are used in salads, the roots to make a coffee-like drink, and the flowerheads are fermented to make wine.
8. Earthshine can be seen close to a new moon, when the whole moon is often bathed in a faint light. The cause is sunlight reflected from the earth.
9. The fairy penguin. It inhabits shallow waters around Southern Australia and New Zealand and is 40 cm. (15 inches) in height.
10. 212°F.
11. Animal species are categorised as extinct if they have not been definitely located in the wild for the past fifty years.
12. A sport in which birds of prey are trained to hunt animals and other birds; also known as hawking.
13. An anomalous weather condition which results in major changes in ocean circulation and biological productivity along the coast of Peru. *El Niño* is Spanish for 'the child' and is so named because the phenomenon usually appears around Christmas time.
14. An evergreen tree or shrub growing to 6 metres (20 feet), native to Somaliland. Aromatic resin is obtained from cuts in the bark.
15. This Japanese globe fish (or puffer fish) is only on the menu in special restaurants. Parts of this fish are poisonous and said to cause instant death. Only restaurant staff who have passed an official examination and have a special licence are allowed to prepare it.

Wildlife Quiz

1. If a plant is described as 'ephemeral', what does it mean?
2. What is the name given to the galaxy to which our sun belongs?
3. What is zoonosis?
4. What are filoplumes?
5. How does the bombardier beetle protect itself?
6. What is a clutch?
7. One creature is thought to be responsible for more human fatalities, in tropical Australian waters, than any other. What is it?
8. What does 'ovovivipary' mean?
9. What is a chocolate-tip?
10. Kidney vetch is a plant that is particularly associated with one of our rarer butterflies. Which species?
11. What small river bird is a good indicator of the health of a river or stream?
12. What colour are the fruit of privet?
13. What fish played its part in the successful completion of the Panama Canel?
14. What is 'bloody Henry'?
15. How many cubs does a red fox normally produce?

ANSWERS

1   It describes an annual plant with a very short life-cycle, usually producing several generations in a single season.
2   The Milky Way.
3   A disease which can be transmitted from animals to humans.
4   Feathers that are like fine hairs which occasionally have a tuft on the ends.
5   When provoked, this beetle fires a caustic cocktail at its attacker. It does so by means of a swivelling 'gun turret' in its abdomen that throws out 50 jets of irritant vapour, each with an explosive click, and at boiling point.
6   A set of eggs, usually the complete number laid by one female.
7   *Chironex fleckeri*, a transparent and almost invisible box jellyfish.
8   The method of reproduction in which the young develop from eggs that are retained inside the body of the mother. Many insect groups, fish and reptiles reproduce in this way.
9   A widespread, locally common moth that is usually associated with woodland.
10 The small blue butterfly which lays its eggs in the flower bud.
11 The dipper. It is particular in choosing clean waters in which to feed and breed.
12 Black.
13 The mosquito fish. This voracious fish devours huge quantities of aquatic mosquito larvae greatly reducing the risk to man of contracting malaria and yellow fever and has been introduced into many areas for this reason.
14 The nickname of the scarlet starfish, which is found on rocky shores around Ireland and Britain.
15 Four or five which are normally born in the spring.

Wildlife Quiz

1. What animal fur was used in Davy Crocket's hat?
2. What is a pangolin?
3. What is an epiphyte?
4. What makes a rattlesnake rattle?
5. What colour are the flowers of the hop plant?
6. Polar bears are found in the Arctic and Antarctic. True or false?
7. Locusts are known as major pest species in some parts of the world. How much can a single swarm consume in a day?
8. What is the bird behaviour known as 'anting'?
9. Can mosquitoes hear?
10. What tree species are favoured by the parasitic plant mistletoe?
11. What is a skink?
12. What is a scat?
13. Where would you find the rostrum on a weevil?
14. What does the adult pale tussock moth feed on?
15. What are the three most conspicuous features that distinguish birds from all other animals?

ANSWERS

1   A racoon.
2   A scaly anteater.
3   A plant not rooted in soil, but growing above ground level, usually on other plants. It uses such plants for support only and is not a parasite.
4   Dead skin on its tail.
5   Green.
6   False. They are only found in the Arctic. There are no land-living native mammals in the Antarctic.
7   A single swarm can number 50,000 million locusts, cover 1,000 square kms. (400 square miles), and can eat 3,000 tons of vegetation in one day.
8   Some 200 bird species are known to use ants to help clean their plumage. They either sit on ant nests or pick up the ants and place them in their feathers. The ants squirt formic acid as a defence and it is believed this helps keep the birds free of parasites.
9   Yes. The female produces an incessant drone by beating her wings at up to 600 times a second. The male's antennae vibrate in sympathy.
10 Mainly apple and poplar trees.
11 A small or medium-sized lizard which has very smooth shiny scales.
12 Another name for animal droppings (dung).
13 It is the weevil's prominent snout which has the antennae attached half way along and the jaws at the end.
14 Nothing. This moth, locally known as the 'hop-dog', (because the caterpillar is found regularly in hop fields) does not feed at all in its adult form.
15 All birds have a beak (bill), a pair of wings, and feathers.

Wildlife Quiz

1. What is eustasy?
2. Maize, barley, wheat, oats and Yorkshire fog are all types of what?
3. Cuttlefish use a form of jet propulsion to move. True or false?
4. Meteorology is the study of what?
5. How many teeth does a blue whale have?
6. What is the longest snake ever recorded?
7. What is the oldest tortoise on record?
8. What is the most numerous family of insectivores?
9. Cat and dog families have one particularly striking similarity between them. What is it?
10. Black-tailed skimmers are found in Ireland. What are they?
11. Spiderlings have been found 'floating' at 5,000 metres (16,000 feet). True or false?
12. The term 'sea-swallow' is sometimes applied to what group of birds?
13. What colour is an alpine chough's bill?
14. What does the word 'eutrophic' mean?
15. Where would you find 'Irish moss'?

ANSWERS

1. It is the term for world-wide changes in sea-levels caused by the advance or recession of the polar ice caps. This has caused a gradual rise in sea-levels over the last century.
2. They are all grasses.
3. True.
4. The weather.
5. None, it has baleen plates and sieves its food.
6. A reticulated python which measured an amazing 10 metres (32 feet 6 inches) in length.
7. A tortoise, which was presented by Captain Cook to the Royal Family of Tonga, in 1773 or 1774. It stayed with the family until it died in 1965 making it at least 188 years old.
8. The shrew family, with around 265 species world-wide.
9. There are 37 living species of cat and 37 species of dogs and their relatives in the world.
10. They are a dragonfly species.
11. True. When spiderlings hatch from their cocoon they head to the nearest high spot – a blade of grass or other vantage spot, and then release a silken thread into the air. This acts as a parachute and assists the spiderlings dispersal by allowing them to drift on the wind, often covering considerable distances.
12. Terns.
13. Yellow.
14. Fertile.
15. 'Irish moss' is another name for carrageen, a small red seaweed often found under other seaweeds in rock pools.

Wildlife Quiz

1. What is the main food of the wombat?
2. What is edaphology?
3. How many weevil species have been identified in the world to date?
4. What kind of tree does turpentine come from?
5. How many kiwi species are there in the world?
6. The flamingo's pink colour comes from the shrimps and algae they eat. True or false?
7. What is Europe's rarest goose?
8. Feral populations of ruddy duck are present in Ireland and Britain. Where did this species originate?
9. What major feature is common to all 800 species of oak?
10. What is the largest ecosystem on earth?
11. What colour is the flower of ragged robin?
12. How many young do hedgehogs produce every year?
13. Which three groups of mammals took to the sea in the course of their evolution?
14. What is a gadwall?
15. The largest order of mammals in the world consists of 28 families of rodent species. How many species are there in total?

ANSWERS

1 Grasses.

2 The study of soil as a medium for the growth of living organisms.

3 Approximately 40,000.

4 A pine tree.

5 Just three; the brown, great spotted and little spotted are all endemic to New Zealand.

6 True.

7 The Greenland white-fronted goose, with just over 30,000 birds world-wide.

8 North America.

9 They all bear acorns.

10 Water (the earth's oceans).

11 Pink.

12 Hedgehogs have between one and two litters each year with four to five young in each litter.

13 Pinnipeds (seals and walruses), cetaceans (whales and dolphins) and sirenians (manatees and dugongs).

14 A duck species.

15 There are 1,650 species in the 28 families. This group accounts for about 40% of all known mammal species.

Amazing Facts

Aged Tree
The oldest living thing in the world to-day has to be a bristlecone pine tree. One such tree, cut down in 1964 for scientific study, was found to have 4,900 rings. A bristlecone found at a height of 9,000 feet in the White Mountains of California is called Methuselah and is estimated to be 4,600 years old. It's thought that bristlecones can live for 5,500 years.

Noisy Insect
By pulsating his tymbal organs at 7,400 times a minute the male cicada can produce a noise so loud that it can be heard 400 metres away. A group of cicadas pulsating together can create a noise louder than a 747 Jumbo Jet and the noise can cause deafness.

Micro-Mite
The male bloodsucking banded louse and a particular parasitic wasp are so small that it would need 1.6 billion of these creatures to weigh just one gram (one insect weighs just 0.005 mg.).

Greedy Gut
The larva of the polyphemus moth consumes 86,000 times its own birthweight in its first 56 days. This is equivalent to a 3.17 kg. baby (7 lb.) taking in 273 tonnes of nourishment.

Desert Maker
The desert locust can eat its own weight in food every day. A swarm of 50 million specimens (this is only a small swarm) can eat food that would sustain 500 people for a year.

Taste A Smell
The male Emperor moth can detect the pheromones of a virgin female of his own species from a distance of 11 km. (7 miles). The chemoreceptors on the male's antennae can detect a single molecule of the scent of which the female carries less than 0.0001 mg.

Land and Water Quiz

1. Which is the longest river in the world?
2. What is the world's strangest substance?
3. What is the world's smallest ocean?
4. Which is the largest lake in Ireland?
5. How big was the largest hailstone ever recorded?
6. Where is the world's largest meteor crater?
7. What is the world's highest waterfall?
8. What is the world's longest reef?
9. What is the world's largest ocean?
10. What is permafrost?
11. Where is the world's coldest permanently-inhabited place?
12. Where is the world's longest glacier?
13. Where is the shortest river in the world?
14. Which is the biggest island in the world?
15. Which is Ireland's largest island?

ANSWERS

1. The Amazon in South America. From source to the Para estuary it is about 6,750 km. (4,195 miles).

2. Water. From a scientist's viewpoint it is very strange. Most substances shrink when cooled, water expands. Most substances are denser in solid form, water is less dense as a solid. If ice sank in water, the world would be a very different place.

3. The Arctic Ocean.

4. Lough Neagh in Northern Ireland.

5. Hailstones of 1 kg. (2 lb. 3 oz.) fell in Bangledash in 1986, killing a number of people.

6. The Barringer Crater in Arizona. It is 1,265 metres (4,150 feet) in diameter and 175 metres (575 feet) in depth.

7. The Salto Angel in Venezuela. The longest single drop is 807 metres (2,648 feet), and it has total drop of 979 metres (3,212 feet).

8. The Great Barrier Reef of Australia. It actually consists of thousands of separate reefs and measures 2,027 km. (1,260 miles) long.

9. The Pacific.

10. Ground that is always frozen solid.

11. The village of Oymyakon in Siberia, Russia. A temperature of −68°C was recorded in 1933 and a recent unofficial figure of −72°C is claimed.

12. In Antarctica. The Lambert Glacier is at least 700 km. (440 miles) long.

13. The North Fork Roe River in Montana U.S.A. It is just 17.7 metres (57 feet 7 inches) long.

14. Greenland.

15. Achill Island, Co. Mayo.

Wildlife Quiz

1. What is neutralism?
2. What is the name of the sun's outer atmosphere which is only visible during a total eclipse?
3. What is, and where would you find, 'dead men's fingers?
4. What is a scaup?
5. What is an amphibian?
6. What is a 'cat's ear'?
7. Where would you find vampire bats?
8. What is a meerkat?
9. What are echinoderms?
10. What is Savannah?
11. What are marsupials?
12. What is a wallaby?
13. Why is the barking spider, of Australia, so named?
14. What is another name for the African honey badger?
15. What family is the skunk a member of?

ANSWERS

1. Where two species co-exist with neither being affected by the association of the other.
2. The corona.
3. A soft coral found in the lower shore region, in rock pools, under pier piles etc.
4. A species of sea duck.
5. Species that include frogs, newts, salamanders and toads. They all have a period of development in water during the tadpole (larval) stage. Their skin tends to be soft and non-scaly.
6. It is a common perennial plant, found in fields, heaths, and roadsides throughout Ireland.
7. South America.
8. A kind of social mongoose found in Africa.
9. Sea creatures such as starfish, sea urchins and sea cucumbers.
10. The grassy plains of Africa (which covers over a quarter of the African continent).
11. Pouched animals.
12. A small member of the kangaroo family.
13. Because it produces a loud noise by rubbing its mouth parts together.
14. The African ratel.
15. The weasel family.

Wildlife Quiz

1. What family is the racoon related to?
2. What is a quill?
3. What is another name for the black-footed penguin?
4. How many species of butterfly are there in the world. Is it 1,200; 5,000 or 15,000?
5. The hyena is a member of what family group?
6. How many species of crow are there world-wide. Is it 75, 116 or 573?
7. Name the largest wild feline in the Mediterranean area.
8. What is another name for the brown rat?
9. What is another name for the Egyptian mongoose?
10. What is the largest snake in southern Europe?
11. How many species of bear are there world-wide?
12. What is a quagga?
13. What is the bush veldt?
14. What is a pan?
15. What South American animal is known as the common zorro?

ANSWERS

1 The bear family.

2 The shaft of a feather.

3 The jackass penguin.

4 15,000.

5 The mongoose family.

6 There are 116 members of the crow family.

7 The Spanish lynx.

8 The Norway rat.

9 The ichneumon.

10 The montpellier snake. It may grow up to 250 cm. (8 feet 4 inches).

11 There are seven species: grizzly, black, polar, sun, sloth, black panda and the red panda.

12 An extinct species of zebra (the last quagga died in a zoo in Amsterdam on 12th August 1883).

13 A common mixed woodland habitat in Africa.

14 A temporary wetland in Africa.

15 The crab-eating fox.

Wildlife Quiz

1. Grey, night, squacco, purple, goliath and black-headed are all members of what bird family?
2. Name the five species of lynx in the world?
3. What is a storm petrel?
4. What is a gallinule?
5. What is a bustard?
6. 'Jackie hangman', 'murdering bird', 'strangler' and 'butcher bird' are all nicknames for what bird family?
7. Green, black leathery, olive ridley, kemps ridley, hawk's-bill, flatback and loggerhead are all members of what family?
8. What is the largest antelope in the world?
9. What is a yucca?
10. What is a sidewinder?
11. Dinosaurs were reptiles. True or false?
12. The name waterfowl is given to which families of birds?
13. Where did the turkey originate?
14. What was one of the first animals to be domesticated by man?
15. What creatures were the first on earth to have a backbone?

ANSWERS

1 The heron family.

2 Canada lynx, Spanish lynx, European lynx, bobcat and caracal.

3 A small swallow-sized sea bird found around our coasts, numerous, but rarely seen.

4 A large colourful moorhen-like bird.

5 A large bird found in parts of Europe and Africa. It is capable of flight.

6 The shrikes. They get their nicknames from the habit of impaling prey on thorns and twigs in open-air larders.

7 They are all sea turtles.

8 The eland, which stands at two metres high (6 feet 6 inches) and is 3.4 metres (11 feet) in length. They are found in grassland areas of Central and Southern Africa.

9 A desert lily.

10 A snake found in desert regions.

11 True.

12 Swans, geese and ducks.

13 North America.

14 The goat.

15 Fish.

Wildlife Quiz

1. What is the smallest antelope in the world?
2. What is a young eel called?
3. How many species of fly are there world-wide? Are there 10,000, 90,000 or 200,000?
4. What is the popular name for the North American woodchuck?
5. What is a jerboa?
6. What does *Tyrannosaurus Rex* mean?
7. What bird family does the condor belong to?
8. How many deer species are there in the world? Is it 29 species, 36 species or 89 species?
9. What is a toucan?
10. What is a palaeontologist?
11. What is torpor?
12. A water violet is a member of what flower family?
13. What is the Irish word for 'wren'?
14. What bird was immortalised in a poem by the English poet John Keats?
15. What is a tuatara?

ANSWERS

1. The royal antelope, which is 25 cm. (10 inches) high, and found in West Africa.
2. An elver.
3. 90,000.
4. The ground hog.
5. A small rodent with long back legs, found in desert regions.
6. King tyrant lizard.
7. The vulture family.
8. There are 36 species.
9. They are middle-sized to large birds with a huge, often brightly-coloured bill. All 38 species are found in tropical forests of South America.
10. A person who studies fossils.
11. A method by which some animals and birds save energy, by lowering their body temperature and slowing down their heart beat.
12. It belongs to the primrose family. Rare in Ireland, it was recently reported found as a native plant in Co. Down and Co. Fermanagh.
13. Dreolín.
14. The nightingale.
15. A type of lizard found only on the North Island of New Zealand. It is the only remaining member of an ancient reptile family, which mostly died out 65 million years ago.

Wildlife Quiz

1. What was a *Pterosaurus*?
2. What is an oxpecker?
3. What was an eohippus?
4. What type of animal is a dorcas?
5. When did dinosaurs become extinct?
6. What is a paradise tanager?
7. Are all piranhas flesh-eaters?
8. The Hindu God, Ganesh, has the head of which animal?
9. What is the unusual feature of how dolphins, whales, and seals sleep?
10. Which bird is the emblem of The Blood Transfusion Service Board?
11. Can bees see the colour red?
12. Baby elephants suck their trunks like human babies suck their thumbs. True or false?
13. How long is a giraffe's tongue?
14. There is a Siamese fish as well as a Siamese cat. True or false?
15. Leaf-cutter, wood and red are all types of what?

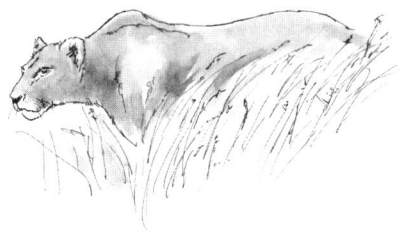

ANSWERS

1. A large flying dinosaur.
2. There are two of these bird species, both found in Africa. Both are closely associated with animals, such as rhinoceroses, feeding on the ticks and other irritating insects that affect these large creatures.
3. The first type of horse. It lived on the earth 50 million years ago and was only 60 cm. (2 feet) high.
4. A gazelle found in North Africa, the Middle East and in India.
5. About 65 million years ago.
6. A particularly attractive and brightly-coloured bird of tropical rainforests.
7. No. Although piranha have a fearsome reputation, some species of this fish feed exclusively on fruit.
8. The elephant.
9. In all these species only half of the brain is asleep. The other half remains awake in case of any threat.
10. The pelican.
11. No, it just looks black to them.
12. True.
13. About 7 cm. (18 inches) in length.
14. True. The Siamese fighting fish which, when danger threatens, alerts the young by rippling his fins and the young take refuge in his mouth.
15. Ants.

Wildlife Quiz

1. To which animal is the okapi related?
2. What is now the most widely accepted explanation for the extinction of the dinosaurs?
3. How many hours a day does the sloth spend sleeping?
4. In the world of cetaceans where would you find a melon?
5. Elephants can detect water underground. True or false?
6. What is a bitterling?
7. Where do bitterlings lay their eggs?
8. What is a marmoset?
9. The arctic fox has fur on the soles of its feet. True or false?
10. The racoon diligently washes its food before eating it. True or false?
11. What is an oropendola?
12. The ancient Egyptian God, Anubai, had the head of what animal?
13. Where would you find trichobothria hairs?
14. How does the female alligator prepare a site for her eggs?
15. Which animal was originally thought to be a cross between a camel and a leopard?

ANSWERS

1. The giraffe.
2. The massive collision between the earth and a celestial object, possibly an asteroid, which hit the earth somewhere near the Yucatan Peninsula, Mexico. The resulting dust cloud is thought to have caused ecological havoc on the planet.
3. 19 hours a day.
4. It is the bulbous forehead of many whales, dolphins and porpoises, believed to focus sounds for echolocation.
5. True.
6. A small silvery fish found in lakes and slow-moving rivers.
7. Inside freshwater mussels.
8. A small monkey, about the size of a squirrel, which lives in South American rainforests.
9. True. The fur grows in winter to prevent the fox from getting frostbite.
10. True. Even if the water is dirty it still washes its food.
11. A bird, about the size of a crow, which lives in South and Central America.
12. A jackal.
13. On spiders' legs. They are specialist hairs, which detect any change in the air.
14. She raises a pile of mud and vegetation with a few sweeps of her powerful tail.
15. A giraffe.

Wildlife Quiz

1. Ray's bream is a fish abundant off the Portuguese coast, but in certain years it makes its way to Irish coastal waters. Who is it named after?
2. Which mammal in the British Isles has the most teeth?
3. In the past why were rowan trees planted in graveyards?
4. Why are grebes, like some other water birds, clumsy on land but adept in water?
5. Sea holly, which is found on most coastal dunes, is a member of which family?
6. Maple syrup is collected from the hive of the maple bee. True or false?
7. Why is the painter's mussel so called?
8. What is maiden's hair?
9. Noctuid moths have ears at the rear of their bodies. True or false?
10. What are nematodes?
11. What tree is the shillelagh made from?
12. How much vegetation does a rabbit need to eat daily, to keep in good health?
13. What colour is meadow saffron?
14. The silver Y moth survives the winter in Ireland by hibernating in adult form. True or false?
15. Where will you find a yellow flag?

ANSWERS

1. It is named after the man who first described it from a specimen found in 1681. He was the great English naturalist John Ray.
2. The mole with 44. All its teeth are sharp pointed; a bite will paralyse a worm, so that it can be stored until the mole is hungry. Moles are unusual in not having proper milk teeth.
3. To keep the dead in their graves and to prevent any ghosts appearing in the neighbourhood.
4. A grebe's feet are positioned right at the back of the body (in the same way that boats have propellers at the stern) to provide good forward propulsion. However this means they have difficulty in moving on land and rarely leave the water.
5. The carrot family. It bears little resemblance to its relatives.
6. False. Maple syrup comes from the sap of the American maple tree.
7. The shell was used by early Dutch painters as a container for their colours. The shell was also considered to be an ideal size for the sale of small quantities of gold and silver leaf used in illustrated manuscripts.
8. One of the brown seaweeds around our coasts and quite often overlooked due to its delicate texture.
9. True. Their ears are very sensitive and are located on the sides of the third (hindmost) thoracic segment.
10. Worms, which are abundant in almost every habitat, including the sea, soil, and the bodies of other animals. Some 20 million of these worms may live under one square metre (1.2 square yards) of grassland.
11. This traditional cudgel is made from the wood of the blackthorn.
12. 500 grams (1.1 lbs.).
13. A delicate pale rosy-pink. This plant is rare in Ireland and can only be found in a few locations in Co. Kilkenny.
14. False. The silver Y moth is a migrant to our shores. This moth does not survive our winter.
15. This tall, yellow-flowered, wetland plant is common throughout Ireland.

Amazing Facts

Small But Deadly
The malarial parasite carried by mosquitoes is probably responsible for 50% of all human deaths, excluding war and accidents, since the Stone Age. In the Sahara 1.4 to 2.8 million people die every year from malaria.

Minute Menace
A virus is so small that it can only be visualised by an electron microscope magnified to about 30,000 times. They vary in size from 0.000018 mm. to 0.0006 mm.

Garden Glut
With unlimited food and no predators, one cabbage aphid could theoretically create an 822 million-tonne mass of descendants every year – more than three times the weight of the world's human population.

Sea Giant
The blue whale, which is almost as long as a 737 aeroplane, consumes 4 tons of krill, sieved from the sea, each day.

Advertising Services
The cleaner, or barber, fish uses its own version of a 'barber's pole', to advertise its services to other fish. It bases itself in a crevice near a brightly-coloured sea anemone or sponge (the barber's pole), and shoals of larger fish queue up to be cleaned of parasites, fungi etc.

Whale Spotting
Unidentified Beaked Whale. So rare that this whale has not yet been given a name. Only 30 positive sightings have ever been made, all at sea, hence the fact that it has never been thoroughly examined and properly named.

Tree Quiz

1. What are broad-leaved trees?
2. What is a mixed woodland?
3. Of these nine tree species which are native to Ireland: beech, sessile oak, willow, birch, alder, hazel, ash, horse chestnut, sycamore?
4. Which tree attracts the most species of insect in Ireland. Is it the oak, beech, or horse chestnut?
5. What tree was native to Ireland but became extinct and was re-introduced around 1652?
6. Which tree is usually used as a Christmas tree?
7. The mountain ash or rowan is native to Ireland. Where did the word 'rowan' come from?
8. Why do trees have leaves?
9. Why do broad-leaved trees shed their leaves?
10. What is the largest tree species in the world?
11. What is a tropical rainforest popularly known as?
12. Why are tropical rainforests so important?
13. Why are wood ants protected in some European countries?
14. How many species of oak are there in the world?
15. Can you eat the fruit of the mulberry tree?

ANSWERS

1. Trees that lose their leaves in winter (with the exception of the holly tree).
2. A woodland with more than one species of tree.
3. Sessile oak, willow, birch, alder, hazel and ash.
4. Oak with 284 species, followed by beech with 64 species and horse chestnut with 4 species.
5. The Scots pine.
6. The sitka spruce.
7. Derived from the Norse word 'runa' which means charm. The tree was believed to ward off evil.
8. Without leaves trees cannot collect energy from the sun.
9. To conserve water and energy.
10. The giant sequoias. They grow over 80 metres (260 feet) and are found in the Sierra Nevada Mountains of California.
11. A jungle.
12. Because they hold the richest and most varied collection of plants, trees, insects, birds and animals anywhere in the world.
13. Because of their great value in destroying forest pests.
14. There are just over 400 species of oak world-wide.
15. Yes. The black fruits look rather like raspberries and are delicious.

Wildlife Quiz

1. The thicknee is the common name for what rare migrant to Ireland?
2. What is the difference between birch and silver birch?
3. What is a sunstar?
4. What is a sulphur tuft?
5. How did the common storksbill get its name?
6. What is a spur-dog?
7. What is an ungulate?
8. In the animal world, what is a vector?
9. Where would you find the Gobi Desert?
10. What is a tansy?
11. There are no glaciers in Africa. True or false?
12. What fish family does the shad belong to?
13. The Danube river is 2,850 km. (1,770 miles) in length. How many countries does it flow through?
14. What is a sawfly?
15. The dunnock and the house sparrow are names for the same bird. True or false?

ANSWERS

1 The stone curlew.
2 The branch ends always hang down in the silver birch whereas they point straight or upwards in birch.
3 A species of starfish, found on rocky shores, usually near the low tide level. This species has 12 arms and looks just like a child's drawing of a sun.
4 A woodland toadstool. Often found in a clump at the base of a tree stump.
5 This small plant, of the geranium family, gets its name from its long twisted seed pod, which looks like a stork's beak.
6 A small shark that rarely grows over 1 metre (3 feet 3 inches). It gets its name from the sharp spine in front of each dorsal fin.
7 A mammal with hooves. There are over 200 species, nearly all are herbivorous with the exceptions being pigs and peccaries.
8 A species, such as an insect, that carries a disease or parasite from one animal or plant to another.
9 In Mongolia and China.
10 A medium to tall plant found in waste areas. It is a strong-smelling member of the daisy family with clusters of yellow flowers.
11 False. Glaciers are found in East Africa on Mount Kenya, Kilimanjaro and in the Ruwenzori Mountains.
12 A member of the herring family.
13 It flows through eight countries. Germany, Austria, Slovakia, Hungary, Yugoslavia (Serbia), Romania, Bulgaria, and Ukraine. It flows into the Black Sea.
14 A group of insects related to bees and wasps. They get their name from the females' saw-like cutting edge of the ovipositor.
15 False. The dunnock's alternative name is the hedge sparrow, not house sparrow. Hedge sparrow is not a particularly suitable name as the dunnock is actually a member of the insect-eating accentor family.

Wildlife Quiz

1. What is a cep?
2. What does the word 'molluscs' mean?
3. Man, bee and monkey are members of what plant group?
4. What is a sea-weed runner?
5. Cinnabar caterpillars are poisonous. What well known migrant bird eats them without harm?
6. The blackthorn flowers appear after the leaves. True or false?
7. What is the other common name for the monarch butterfly?
8. What colour is the inside of the mouth of a raven's nestling?
9. What does sessile mean?
10. What is the smallest mammal occurring in Ireland and Britain?
11. The robin is a member of the thrush family. True or false?
12. A spotted redshank's legs are normally red. What colour do they change to when the bird is in summer plumage?
13. The common seal gives birth to its young in water. True or false?
14. How did the kittiwake get its name?
15. Only one tree is planted for every ten that are cut down. At this rate, how soon will all the remaining tropical forests disappear?

ANSWERS

1 A widespread edible fungus found in deciduous woodland.

2 'Soft', meaning soft fleshy-bodied animals.

3 They are all orchids.

4 A fly species found amongst decaying seaweed on the sea shore.

5 The cuckoo.

6 False. It flowers before the leaves arrive, unlike the hawthorn (whitethorn) which is the other way round.

7 It is sometimes called after its larva foodplant which is the milkweed.

8 Bright red. Quite startling as the bird is black in colour.

9 Without a stalk.

10 The pygmy shrew.

11 True.

12 Black.

13 False. They give birth to their young on a rock or sandbank. The young are able to swim from birth.

14 From its call which is a loud *kittee-wake, kittee-wake*, heard from their breeding colonies on sea cliffs.

15 At this rate it is estimated that these forests, which today cover 12% of the earth's land surface, will be destroyed by the year 2035.

Wildlife Quiz

1. What colour are ripe elderberries?
2. Certain tiny midges have a wing-beat rate of 50,000 beats a minute. True or false?
3. What is escallonia?
4. In the nursery rhyme, what frightened Miss Muffet away?
5. What tree is a hurley made from?
6. In which months does the ivy flower?
7. The electric eel of the Amazon river can produce 1,000 volts of natural electricity. True or false?
8. What colour are the legs of a sparrowhawk?
9. What is the largest invertebrate known in the world?
10. Where does the word 'zoology' come from?
11. Wild bananas are mainly pollinated by what?
12. There are over 275,000 flowering plants and trees presently known to science. True or false?
13. What is a Tibetan pika?
14. What colour are the legs of an adult lesser black-backed gull?
15. What does 'abiotic' mean?

ANSWERS

1. Black.

2. True. It is a staggering figure of over 830 wing-beats a second.

3. A shrub seen growing wild as well as in garden hedges. Leaves oval and fleshy, flowers pink. It is an introduced plant and not native to Ireland or Britain.

4. A big spider.

5. Ash.

6. September to November.

7. False. It produces natural electricity up to 500 volts and not as stated.

8. Yellow.

9. The giant squid, which can exceed 18.5 metres long (60 feet).

10. It is from the Greek word zoion which means animal.

11. Bats, as they hunt for nectar during the night.

12. True.

13. A mammal related to the rabbit and occupies one of the highest habitats of any animal as it lives in the Himalayas at up to 6,000 metres (20,000 feet).

14. Yellow.

15. Devoid of life. Non-living. The opposite to biotic which means the living.

Wildlife Quiz

1. Why does a robin perch on a spade?
2. What is a Miller's thumb?
3. What medicinal properties is watercress believed to possess?
4. What colour is a choughs' bill?
5. Which animal spread the bubonic plague in the middle ages?
6. What is a young pike called?
7. What tree is also known as 'golden chains'?
8. How many eggs does a puffin lay?
9. What is the difference between a common hedgehog and a wood hedgehog?
10. What bird species never sets foot on land during its life?
11. What is the difference between a foxglove and a foxglove pug?
12. What bird species have asymmetrical ears?
13. What is hartstongue?
14. What is sea ivory?
15. What bird was deliberately burnt by order of the Vice-Chancellor of Oxford University?

ANSWERS

1. Robins like to use a low perch to survey the area. A spade is an ideal low perch. From these vantage points they can fly down to take their prey.
2. A small, spiny, fresh-water fish, found in England and Wales.
3. Believed to aid blood purification. It is also used to help relieve headaches.
4. Red.
5. The black rat (its fleas carried the disease). This epidemic, known as the Black Death, wiped out more than a quarter of the population of Europe during the 14th century.
6. A jack.
7. The laburnum. Because of its spectacular display of pendulous yellow flowers.
8. One, rarely two.
9. The common hedgehog is a mammal and the wood hedgehog is a type of mushroom.
10. The Emperor penguin. It lives and breeds on the Antarctic sea ice.
11. The foxglove is a plant and the foxglove pug is a moth.
12. Owls.
13. A fern of damp and shady woods and banks.
14. A lichen found on coastal rocks and stones. It is tolerant of sea spray.
15. A dodo (it was the only stuffed specimen in existence at the time and was burnt in the 18th century as it was getting moth-eaten).

Wildlife Quiz

1. The male glow worm does not glow in the dark. True or false?
2. What colour is a nightingale's egg?
3. How does a click beetle get its name?
4. The slowest plant to flower in the world is a *Puia raimorid* from Peru. How many years does it take for the central cluster of the flower-head to emerge?
5. Is there such a bird as a 'willy wagtail'?
6. Where would you find a black robin?
7. What colour are soldier beetles?
8. What is a logrunner?
9. In relation to birds, what do Cape Clear Island, Co. Cork and Copland Island, Co. Down have in common?
10. *Zostera* species are dark green grass-like plants that grow in estuaries and creeks, and are one of the main food plants of brent geese and many duck species. What is this plant's common name?
11. Where would you find the sickener?
12. What is allogrooming?
13. Where would you find a dickcissel?
14. Male crossbills are predominantly a reddish colour. What colour is the female?
15. What is particularly distinctive about the male narwhal, and where would you find one?

ANSWERS

1 True. It is the female that produces a cold light from her abdomen to attract a passing male.
2 Dark brown. They can vary from greyish-green to reddish-buff.
3 If caught on its back this insect has the ability to turn back over by flicking itself into the air, making a loud click in the process.
4 150 years.
5 Yes. A common bird in Australia which is actually a fantail species and not a wagtail species.
6 In New Zealand. They are just like our robin except they are black.
7 Orange-red. They are also known as sailor beetles. Their names come from their bright colour which is reminiscent of military uniforms.
8 It is a bird of Australian rainforests and feeds on the forest floor.
9 Both islands have active Bird Observatories operating on them.
10 Eel-grass (occasionally sea-grass).
11 It is a widespread and colourful fungus found in conifer woodlands and, as the name suggests, is poisonous.
12 Where one animal grooms another of the same species (autogrooming is where the animal grooms itself).
13 A sparrow-like bird found in grain fields and weedy patches in North America.
14 Yellowish grey-green.
15 Male narwhals have a single tusk which protrudes from the mouth forwards for about half the total length of the body. They are found in the Arctic, usually up in the Arctic circle, and have one of the most northerly ranges of any whale.

Wildlife Quiz

1. What is a common predator of the limpet?
2. What do the letters S.A.C. stand for?
3. What crab risks drowning when breeding?
4. What colour are the flowers of sea holly?
5. Does the North American opossum really 'play possum'?
6. How big can a giant clam grow?
7. When did the Irish elk become extinct?
8. Why is a male swallow's tail important?
9. How long can a tropical bird-eating spider live?
10. Mayflies are credited as being the shortest-lived insects. How long do they live as adults?
11. Why is a pistol shrimp so called?
12. What is so special about a cormorant's eyes?
13. What colour are the berries on a juniper tree?
14. Why is a barn owl's face heart-shaped?
15. What colour is a black woodpecker's crown?

ANSWERS

1. The dog whelk. It bores a hole through the shell and sucks out the soft body inside.
2. Special Areas of Conservation. Habitats of special importance protected under European Union legislation.
3. The red land crabs of Christmas Island in the Indian ocean. 120 million red land crabs live on the forest floor, breathe air, and cannot swim. Each year they have to return to the sea shore as their larvae can only develop in the sea.
4. Powder-blue.
5. Yes. This marsupial feigns death when threatened. It lies completely still, eyes open, yet is fully alert. It can remain in this state for hours if necessary.
6. The giant clam shell can be up to one metre across (3 feet), and can weigh up to 180 kg. (400 lbs.).
7. About 13,000 years ago. Due to the massive size of its antlers (up to 3.6 metres/12 feet across), it was unable to move through Ireland's developing forests.
8. The length and symmetry of a male's tail indicate a strong, fit bird. Females select longer-tailed mates for this reason.
9. They can live for up to 25 years.
10. They live for as little as one hour as an adult. The nymph stage, spent at the bottom of a lake or stream, lasts from two to three years.
11. Because of the pistol-like shot of sound it makes clicking its huge claw. The sound stuns its prey and is so loud it can actually shatter glass.
12. Its eyes have soft lenses which can be changed to allow focusing at greater depths.
13. Dark blue.
14. The shape helps focus sounds directly to its ears.
15. Bright red.

Wildlife Quiz

1. In Antarctic waters, what hazard is known as a 'growler'?
2. What are ramsons?
3. What is a cold-blooded creature?
4. The Law of the Sea is a branch of International Law and divides the sea into three zones. What are these zones?
5. Which plant, the subject of many superstitions, has been claimed to scream when uprooted?
6. What worm likes to use a 'mobile home'?
7. How did the plant rest-harrow get its name?
8. What bird makes the world's biggest nest?
9. What is the speculum of a duck?
10. What is the only bird known to hibernate?
11. What is the world's biggest burrowing animal?
12. What is a 'true lover's knot'?
13. What colour is the fruit of a bilberry?
14. What does 'deciduous' mean?
15. What large bird is regarded as a good omen in many parts of Europe?

ANSWERS

1. A small, and usually difficult to see, piece of ice (from an iceberg), awash with waves and thus a hazard to shipping.
2. Wild garlic.
3. A creature that cannot regulate its body temperature, e.g. reptiles, fish and amphibians.
4. Internal Waters – includes ports, rivers, lakes and canals. Territorial Waters – the width of sea adjacent to a state and legally owned by that state. Traditionally 4.8 km. (3 miles). The High Seas – outside territorial waters and may be used freely by all shipping.
5. The mandrake. This thick-rooted perennial, native to Europe, produces blue flowers and yellow-orange berries and was once widely regarded for its medicinal and narcotic properties.
6. The ragworm. Hermit crabs take over empty shells, usually bigger than necessary to allow for growth and this extra space is sometimes taken up by a ragworm which moves in as a room mate.
7. When farmers depended on horse-drawn ploughs and harrows, this particular plant used to become entangled causing the farmer to 'rest' the 'harrow' in order to free it.
8. The weaver bird. These birds build an enormous communal nest containing up to 300 woven grass baskets all clustered under one dome-shaped, thatched roof. The structure may be 4 metres (13 feet) tall and 14.8 metres (48 feet) across.
9. A patch of distinctive colour on the wing.
10. The common poorwill of North America. In winter this species hibernates in rock niches and may return to the same hole every winter.
11. The wombat of South-East Australia. This animal is about as heavy as an Old English sheepdog.
12. A moth, which is found in Ireland.
13. When ripe, a purplish colour.
14. A woody plant that sheds its leaves in winter.
15. The white stork. These birds are so highly regarded that many homes have a permanent, protected stork's nest on the roof. Storks return to the same nest year after year and one nest on a tower in Eastern Germany is recorded as first being occupied in 1549, and was still in use in 1930.

Amazing Facts

Whale Tale
The skeleton of Old Tom, a killer whale, is housed in the museum at Eden, a small township in New South Wales. This remarkable whale, with its companions, used to join up with the local whalermen and helped them to trap and kill humpback and/or fin whales. Old Tom began his partnership with man in 1843 and the relationship continued to his death in 1930. Even when there were no boats at sea, the killer pack would surround a whale, then send two or three of their number to raise the alarm in Eden by slapping the sea with their flukes. The whalermen would then follow the killers and assist in the capture.

Einstein Of The Sea
The largest brain in the world belongs to the sperm whale. Its brain weighs more than 9 kg. (20 lbs.). It also happens to be the deepest-diving sea mammal of all. It can swim down to depths of over 1 km. (3,300 feet).

Whale Of A Song
Each male humpback whale has its own song that may last up to 35 minutes.

Tuning In
The frog-eating bat of Central America can tell from the sound of a frog whether it is poisonous or edible.

Royal Compass
Monarch butterflies migrate 3,000 km. (1,875 miles) using the earth's magnetic lines.

Banana Bunch
Over 40 million tonnes of bananas are eaten each year making it the world's most popular fruit (to humans!).

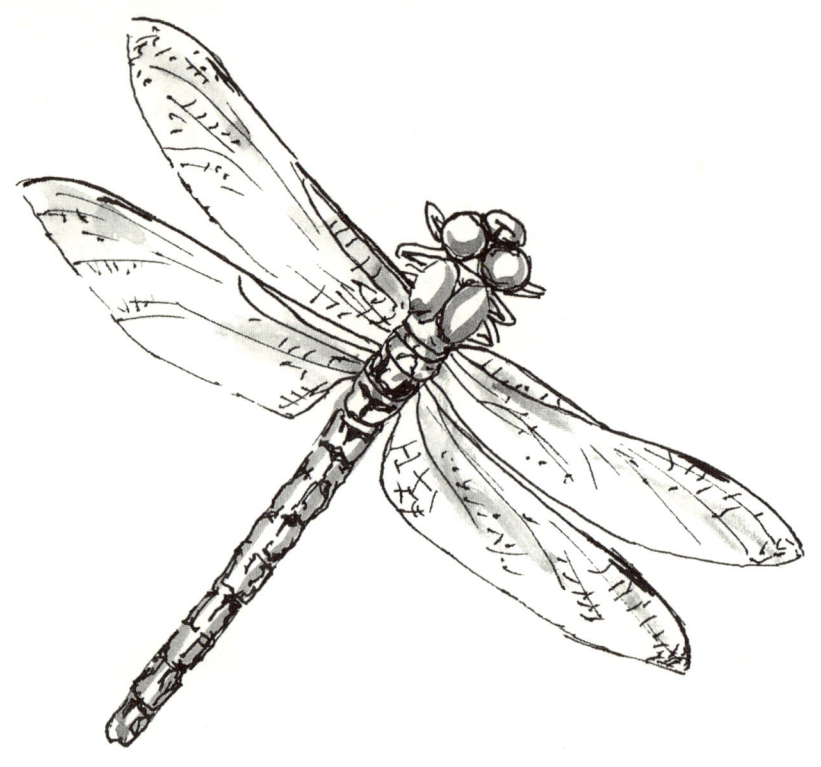

Insect Quiz

1. What is a flea?
2. What is a cocktail?
3. What is the ovipositor?
4. What is the leathery forewing of cockroaches, grasshoppers and related invertebrates called?
5. How many known species of true flies (order *Diptera*) are there in the world. Is it 10,000; 50,000; 100,000 or 500,000?
6. What is the difference between a butterfly and a moth?
7. What insects did Moses use as one of the plagues on the Egyptians?
8. The ladybird is a beetle. True or false?
9. What does the word 'apterous' mean?
10. What is a daddy-long-legs?
11. Where does the word 'butterfly' come from?
12. What is the association between 'horse-stingers' and 'devil's darning needles'?
13. Ichneumon flies are parasitic. What do they prey on?
14. What is a firebrat?
15. What is a burnished brass?

ANSWERS

1. A small wingless, blood-sucking invertebrate with enlarged hind legs for jumping.
2. Another name for the devil's coach-horse, a member of the rove beetle group. When disturbed they raise their hind end and open the jaws wide in a threatening attitude. They hunt at night for slugs and other invertebrates.
3. The egg-laying organ.
4. Tegmen.
5. The order of true flies is huge with 100,000 known species.
6. The only true difference is that butterflies have club-horned antennae (clubbed), whereas the moth has variously-horned antennae (feathery or pointed).
7. Locusts.
8. True.
9. Wingless.
10. A crane fly. There are many different species, all members of the order *Diptera* (true flies).
11. Brimstone. The word originating from the brimstone's 'butter-colour' (butter-coloured fly).
12. Horse-stingers is an old name for dragonflies and devil's darning needles is an old name for damselflies. They belong to the insect order of *Odonata* (dragonflies).
13. These parasitic insects lay their eggs in other insects. The female lays just one egg in each of its hosts.
14. A bristletail very similar to the silverfish, though browner in colour. It has the three long 'tails' and prefers warm places such as heating ducts and bakeries.
15. A common moth which has metallic-like greenish-gold patterning on its wings. The larvae feed on nettles, burdock and spear thistle.

Nature Quiz

1. What is meant by 'arboreal'?
2. What is nectar?
3. What is the largest rodent in Europe?
4. What colour are the flowers of a rowan tree?
5. What is the largest flower in the world?
6. Which has more teeth, a cat or a dog?
7. Where would you expect to find the 'old lady'?
8. Which bat has the largest wing-span?
9. What is a 'fat hen'?
10. What is the *Aurora borealis*?
11. What is a leaf miner?
12. What does the name hippopotamus mean?
13. Ireland is the only European country where the Killarney fern is found. True or false?
14. What kind of a bird is a phoenix?
15. What is a 'liberty cap'?

ANSWERS

1. Tree-dwelling.
2. A sweet fluid produced by plants.
3. The European beaver.
4. White with yellow centres.
5. The giant rafflesia (*Rafflesia arnoldii*), which can grow up to 105 cm. (3 feet 6 inches) and weigh up to 7 kg. (15 lbs. 5 oz.).
6. A dog, which has 42 teeth whereas a cat only has 30.
7. The old lady is a widespread moth species found in gardens, waste ground, river banks and marshy places. The adult flies in July and August.
8. The fruit bat with a wing span up to 2.0 metres (6 feet 6 inches).
9. A common and widespread annual plant found in waste places and disturbed ground.
10. Commonly known as the Northern Lights. Rarely seen in these islands. The effect is of streamers of light in night skies, usually observed in polar regions.
11. An insect that spends its early life tunnelling between the upper and lower surfaces of leaves. Most leaf miners are the larval form of small fly and moth species.
12. River horse.
13. False. Although rare, it is found in England, Wales, Scotland and in the extreme west of the European Continent. It is considered an Atlantic species.
14. There is no such bird in reality. However, in mythology the phoenix was an Arabian bird, the only one of its kind, said to live for 500 to 600 years in the desert, burn itself on a funeral pyre, and rise again in renewed youth.
15. A small fungus that got its name from its cap [measuring up to 1 cm. (½ inch) across] which apparently resembles the cap worn by peasants during the French Revolution.

Nature Quiz

1. What does an ecologist study?
2. The great skua has a well-known nickname. What is it?
3. How many legs does a crab have?
4. For its size, which bird has the most powerful bill and jaw muscles in the world?
5. What is immunology?
6. What are fescues?
7. What is the more common name for a pupa?
8. What is a common swift?
9. What animal featured in the book and film *Watership Down*?
10. What is a spore?
11. The clouded yellow, painted lady and red admiral all have something in common. What is it?
12. Two common pipit species are present in Ireland. What are they?
13. Where would you find a Tasmanian devil?
14. What is a rudd?
15. The comma butterfly is common in Ireland and Britain. True or false?

ANSWERS

1. The study of relationships amongst and between organisms, and between them and all aspects, living and non-living, of their environment.

2. Bonxie.

3. Ten. Shrimps, prawns, crabs and lobsters all have five pairs of legs.

4. The hawfinch – rare in Ireland, common in Britain. This finch has a heavy bill and huge jaw muscles and is able to cleanly cut a damson stone in half. It is believed that the force exerted is up to 170 kgs. per square cm. (150 lbs. per square inch).

5. The scientific study of immunity.

6. They are species of grass.

7. A chrysalis. The third stage in an insect's life-cycle.

8. Common swift refers both to a bird and a moth. The bird, scientific name *Apus apus*, is a common summer migrant to Ireland from Africa. The moth, scientific name *Hepialus lupulinus*, is a widespread resident species in Ireland.

9. The rabbit.

10. The microscopic structure essential to the reproduction and dispersal of fungi. A spore does not contain an embryo and is therefore distinct from a seed.

11. They are all migrant butterflies from the continent.

12. The meadow pipit and the rock pipit.

13. In Tasmania. It is a ferocious nocturnal marsupial.

14. A freshwater fish found in slow-moving waters.

15. False. Although a common resident in Britain, it has not been recorded in Ireland.

Nature Quiz

1. Where would you find the world's tropical forests?
2. What colour are the flowers of tree lupin?
3. It is estimated that in excess of 175 million tonnes of solid waste is dumped into the world's oceans each year. 80% of that waste comes from one human activity. What is it?
4. What does the West Indian wood snake do when threatened?
5. The game conkers is played with the fruit of a tree. What was the game originally played with and what was it called?
6. How does an Egyptian vulture break open an ostrich egg?
7. What plant is the national emblem of England?
8. How does a remora fish travel?
9. What is a jojoba?
10. How long did dinosaurs rule the earth?
11. Which is the largest lake in the world?
12. What bird species is known as a goatsucker?
13. What is the death-watch beetle famous for?
14. Cuckoo wasps lay their eggs under the skin of birds' nestlings. True or false?
15. The shag, a close relation of the cormorant, is of coastal distribution and breeds in loose colonies. It is not found inland. True or false?

ANSWERS

1. Tropical rainforests are those which lie between the tropic of Capricorn and the tropic of Cancer. They are made up of evergreen trees.
2. Yellow.
3. The dredging of rivers to maintain shipping channels.
4. Not only does it pretend to be dead, it gives off a foul stench, like decomposition, and tiny blood vessels trickle blood from its eyes and mouth.
5. Originally sea shells and the game was then called conches.
6. It picks up a stone with its bill and throws it at the egg.
7. The rose.
8. It attaches itself, with a sucker, to whales, sharks and other large fish and hitches a ride.
9. A desert shrub found in Arizona, California and Mexico.
10. They dominated the earth for over 140 million years (man has only been on the planet for 2 million years).
11. The Caspian Sea, which is actually a lake. It is 371,800 square kms. (143,552 sq. miles) in size and borders Russia, Kazakhstan, Turkmenistan, Azerbaijan and Iran.
12. The nightjar. This species is a rare summer-breeding visitor to Ireland.
13. It is a wood-boring beetle renowned for the serious damage it causes to timber in many older buildings. It gets its name from the sound the adult beetle makes as it taps its head against the wood. This was once believed to foretell a death.
14. False. Parasitic wasp species lay their eggs in the nests of other wasps and bees.
15. True. The shag has a coastal distribution and is, very much, a marine habitat species whereas the cormorant is often found inland along rivers and on inland lakes.

Nature Quiz

1. Where is the pelvic fin on a fish?
2. What is the group name for plant-eating dinosaurs?
3. What is a neap tide?
4. The green sandpiper nests in the north and east of Europe. Where do they normally lay their eggs?
5. What is a decapod?
6. What is the official scientific language used for naming all species?
7. The mallee fowl of Southern Australia spend five months of the year tending their nests. When the chicks hatch, neither parent takes any notice of them. True or false?
8. In ancient times, cowslips were used for particular medicinal purposes. What were these?
9. An asp is a fresh-water fish belonging to the carp family. True or false?
10. How did the barnacle goose get its name?
11. What is a blood worm?
12. What colour is the alder moth caterpillar?
13. Broom is a member of the pea family. True or false?
14. What is a catkin?
15. Burnet moths are brightly coloured because they contain cyanide and are very poisonous. True or false?

ANSWERS

1. The pelvic fins are found on the underside of a fish's body.
2. Sauropods. They were extremely large and heavy and generally quadrupeds.
3. The tide that has the smallest difference in water level between high and low tides.
4. They usually lay their eggs in the old nests of thrushes and woodpigeons in conifer trees.
5. A ten-legged crustacean. That is five pairs of walking legs.
6. Latin.
7. True. Eleven months of the year are taken up laying and tending the nest. The chicks hatch after 50 days in the nest and have to dig their way out of the elaborate nest structure, fend for themselves, and are totally ignored by both parents.
8. A 'tea' made from the petals was used to calm the nerves and bring about a sound sleep.
9. True. The asp is a solitary fish, which can grow up to 1 metre (3 feet 3 inches) long, and inhabits slow-running rivers in Central Europe.
10. It was once thought that this bird hatched from the barnacles on the sea shore. Hence the name.
11. The bright red larvae of certain midges that live in stagnant water.
12. Black and bright yellow. It is also particularly unusual in that it has paddle-shaped black hairs that stick out all along the sides of the body.
13. True.
14. It commonly applies to the male flowers of trees such as hazel, poplar and willow and is often seen hanging in long clusters in the spring.
15. True. These moths are day-flying and the bright colours are to warn potential predators of the danger of eating them.

Nature Quiz

1. Which is the only Irish butterfly to have wings that are really green?
2. What is a polyp?
3. The hedgehog population in Ireland is of international importance. True or false?
4. What common garden 'weed' is chicory related to?
5. What is a dab?
6. Where would you find wild cranberry growing?
7. In ants what is the acidopore used for?
8. What is the other name for a cuckoo flower?
9. What is a hermaphrodite?
10. What colour are the eyes of a shag?
11. What is the more common name for the herb Gerard?
12. What causes a fairy ring?
13. Which of the big cats cannot withdraw its claws into their sheaths?
14. What is another name for a gnu?
15. Why is the lungfish so called?

ANSWERS

1. The green hairstreak. The orange tip butterfly also appears to have green on the wings but this is an illusion created by the black and yellow scales inter-mixed on the underwing.
2. A marine animal which normally consists of a sac-like body with tentacles surrounding its mouth. Jellyfish, sea anemones and corals all belong to the polyp group of animals.
3. True. Although widespread and common in Ireland the population of hedgehogs in Europe is threatened. This makes the Irish population very important.
4. The dandelion. The chicory's leaves can be used in salads.
5. A flat-fish similar to a plaice but smaller, usually found in shallow waters.
6. In bogs and cutovers. The plant is a low, creeping evergreen shrub that has small pink flowers in the spring.
7. It is the small pore some ants have at the tip of the abdomen, from which defensive secretions are released.
8. Lady's smock.
9. An organism that has both male and female reproductive organs (combined in the same individual).
10. Green.
11. Ground elder. During the Roman Empire this plant was used as a vegetable and was also used to treat gout and sciatica.
12. It is caused by various fungi species growing in the soil. At certain times of the year the toadstools grow up through the surface causing a conspicuous circle. As food sources are exhausted the fungal threads keep pushing outwards from the centre creating the rings seen in wood and grassy areas. One large fairy ring was estimated as being over 200 years old.
13. The cheetah.
14. Wildebeest.
15. Unlike most fish it has lungs as well as gills, allowing it to survive out of water for some time.

Nature Quiz

1. Which planet is closest to the sun. Is it Jupiter, Mercury or Venus?
2. Where would you find pepper dulse?
3. Where is the world's largest marine park?
4. Which is the world's largest butterfly?
5. What is the tail of a fox called?
6. Which Irish bird regularly walks under water looking for food?
7. What is the main food plant of the common blue butterfly?
8. What is a rustyback?
9. What is a leatherjacket?
10. What is a reptile?
11. What is a common scoter?
12. The population of common frogs is designated in Ireland as of international importance. True or false?
13. Where would you look for hound's tongue?
14. The rare freshwater pearl mussel, *Margaritifera margaritifera durrovensis* can only be found on one 10 kilometre stretch of river in Ireland. Which river is it?
15. The little tern is one of Ireland's rarest breeding bird species. Where do they lay their eggs?

ANSWERS

1. Mercury.
2. On the seashore. This seaweed has a strong peppery taste – hence the name.
3. The Great Barrier Reef National Park, Australia. It is 207,000 square km. (80,000 square miles). Species seen in the park include humpback whale, whale shark, sea turtle and around 1,500 fish species and 350 coral species.
4. The Queen Alexandra's birdwing found in Papua New Guinea and the species is endangered.
5. A brush.
6. The dipper (an Irish subspecies).
7. Birdsfoot trefoil.
8. A type of fern that likes to grow on walls. It gets its name from the colour of the underside of its leaves.
9. The larva of a crane-fly (daddy-long-legs).
10. They are air-breathing from hatching onwards, with no gilled larval phase. Bodies are covered with ectodermal scales, sometimes supported by bony plates.
11. A sea duck that is one of our rarer breeding species, nesting on islands of inland lakes in the midlands and west of Ireland.
12. True. The common frog is considered as widespread and common in Ireland but is 'vulnerable' in the rest of Europe.
13. This plant likes dry grassy places and sand dunes and is found particularly in the east and south of Ireland. It is rare inland.
14. The River Nore in Co. Waterford. There are probably only about 150 individuals left and as they are not reproducing, this species will become extinct in the 21st century.
15. They usually breed in small colonies, on sand and shingle beaches along the coast. They lay their two to three eggs in a shallow scrape made amongst the stones and shells. The nest is very difficult to see.

Nature Quiz

1. Which Roman Emperor knew there was going to be an eclipse of the sun on his birthday?
2. Is the holm oak deciduous or evergreen?
3. What does 'altricial' mean?
4. What is the main food plant of the cinnabar caterpillar?
5. What is an orange tip?
6. What is the study of animal behaviour called?
7. What colour is the flower of the common twayblade orchid?
8. What is a lek?
9. The common butterwort belongs to a very select group of plants. What makes it unusual?
10. Which member of the crow family is involved in the game of chess?
11. What plant is the national emblem of Wales?
12. What colour is the wood of a strawberry-tree?
13. The jay is a member of the crow family. True or false?
14. What colour are the legs of a black guillemot?
15. What is a muntjac?

ANSWERS

1. Claudius.
2. Evergreen. Oak species contain both evergreen and deciduous species.
3. Another word for nidicolous, which is when a chick hatches from the egg and is naked, blind and totally helpless.
4. Ragwort.
5. A butterfly. The male has orange tips to the wings.
6. Ethology.
7. Green.
8. A communal display ground where male birds of certain species congregate for the sole purpose of attracting and courting females.
9. This perennial plant of bogs and damp places is carnivorous.
10. The rook.
11. The leek.
12. Pink. This hard, fine-grained wood is highly prized by furniture makers.
13. Yes.
14. Red.
15. The smallest deer found in the British Isles (also known as the barking deer and introduced into Britain at the beginning of the 20th century from China and Taiwan).

Amazing Facts

Heads Will Roll
During an eclipse in 2,136 BC in China people thought the sun was being eaten by a dragon. The Emperor of China ordered the Imperial Astronomers beheaded for failing to warn him of the dreadful monster.

Tall Tale
In Eureka, California in 1967, Roger Patterson filmed a giant hairy forest creature, known to native Americans as Bigfoot, or Sasquatch.

Down The Drain
Ever wondered why water goes down the plughole clockwise in the Northern Hemisphere and anticlockwise in the Southern Hemisphere? It's all to do with the rotation of the earth and is called the Coriolis Force.

Seeing Is Believing
Leaf insects not only resemble leaves in colouring and markings, but can flap their bodies imitating a leaf trembling in the wind.

Second Sight
The bamboo viper has deep pits between the eyes that can act as a second sight. These heat sensors help the snake to find warm-blooded prey in total darkness.

Flight Feat
The arctic tern is believed to see more daylight than any other living creature as it breeds around the Arctic, Northern Atlantic and Pacific Oceans, and travels south to winter off the Antarctic pack-ice. They are believed to travel 32,000 km. (20,000 miles) on migration each year. (That is just the journey). Arctic Terns are known to be able to live to at least 34 years, so on migration alone they would be travelling over one million kilometres (625,000 miles) in a lifetime.

Whale Quiz

1. How many species of whale and dolphin are there?
2. What is a whale's tail called?
3. What whale is affectionately known as the 'canary of the sea'?
4. When a whale is 'spy-hopping', what does it mean?
5. What is a whale doing when it is breaching?
6. How many blow holes does a baleen whale have. One or two?
7. Whales, dolphins and porpoises are known collectively as what?
8. Where did the blue whale skeleton, displayed in the Natural History Museum, London, come from?
9. What is a baby whale called?
10. A beluga mother is white, what colour is her calf?
11. How do whales, dolphins and porpoises navigate?
12. In relation to whales, what does the term 'logging' mean?
13. What is another name for the common porpoise?
14. Which dolphin was featured in the TV series *Flipper*?
15. What is the largest whale in the world's oceans?

ANSWERS

1. 79 species.
2. The flukes, which are horizontally positioned and contain no bones.
3. The beluga (because it is very vocal).
4. When a whale is curious and lifts its head out of the water.
5. Leaping out of the water.
6. Two blow holes.
7. Cetaceans.
8. Wexford Harbour (The animal was stranded on 25th March 1891 but was not displayed until 1933).
9. A calf.
10. Grey.
11. By using echolocation, a similar system to bats.
12. When whales are seen floating motionless on the surface of the water and all facing in the same direction.
13. The harbour porpoise or the puffing pig.
14. A bottle-nosed dolphin.
15. The blue whale [up to 33 metres (110 feet)].

Nature Quiz

1. Where and when was the highest density of rodents found in the world?
2. What does the word 'olfactory' relate to?
3. What is the name given to a sea urchin's shell?
4. A coot is a water bird of lakes, reservoirs and large meandering rivers. What colour is its bill?
5. What is frass?
6. 'A little bit of bread and no cheese' describes the song of a familiar farmland bird. Which bird is it?
7. What is a young pilchard called?
8. What is dendrochronology?
9. Hips are the fruit of what plant?
10. 'Loon' is an American word for what bird species?
11. Which is Ireland's smallest butterfly?
12. What are parr?
13. The maple leaf is the nation emblem of which country?
14. What is a weed?
15. What is the name of a beaver's home?

ANSWERS

1 In Kern County, California in 1926 a colony of house mice was discovered that contained 205,000 animals per hectare (83,000 per acre).

2 The sense of smell.

3 The test.

4 A coot's bill is white.

5 Insect droppings.

6 The yellowhammer.

7 A sardine.

8 The analysis of tree rings (ageing the tree is just one part of it). For example, Irish bog oaks, dating back to circa 4,000 BC, have proved a crucial check on radiocarbon dates.

9 Wild roses.

10 Divers. There are two common and two rare species recorded in Ireland.

11 The small blue (also known as the little blue).

12 The juvenile freshwater stage of salmon. They remain in the river for approximately two years before changing into smolt and migrating to the sea.

13 Canada.

14 A misplaced plant.

15 A lodge.

Nature Quiz

1. What is a dabchick more commonly called?
2. What is the fruit of a hawthorn called?
3. What is the substance mumiyo?
4. What is an otter's spraint?
5. What is a common name for the gorse bush?
6. A male falcon is known as what?
7. On what plant does the holly blue butterfly lay its second brood of eggs in the year?
8. What do the following birds have in common: swift, swallow, spotted flycatcher, and willow warbler?
9. What is a bird's nictitating membrane?
10. Bog asphodel is a plant of bogs and wet heaths. What colour are its flowers?
11. DNA is the complex molecule that is a living creature's blueprint to replicate itself. What do the initials DNA stand for?
12. *Araucaria araucana* is the scientific name of a 'difficult to climb' ornamental tree introduced from Chile at the end of the 18th century. What is its common name?
13. Where would you look for purple hairstreaks?
14. Dutch elm disease is caused by a fungus. How is it spread from tree to tree?
15. What is detritus?

ANSWERS

1 Little grebe.
2 Haws.
3 The oily substance regurgitated as a defence mechanism by certain seabirds. In the Antarctic deposits of this substance, which have built up at certain breeding colonies over thousands of years, have been aged by radio-carbon-dating and found to go back as far as 34,000 years.
4 Its faeces (droppings).
5 Furze bush.
6 A tiercel.
7 On ivy. The first brood is laid mainly on holly. Second broods in Ireland are relatively rare, with records in many of the southern counties.
8 They are all summer migrants to Ireland, wintering in Africa.
9 A transparent fold of skin which can be drawn over a bird's eye as a third eyelid.
10 Orange-yellow. The previous year's dried plants are often found in the same area.
11 Deoxyribonucleic acid. No wonder people prefer to use the initials instead!
12 Monkey-puzzle.
13 These butterflies are found high up in the canopy of oak trees and rarely found anywhere else. Distribution in Ireland is confined to oak woods in the southern half of the country, though it does not appear ever to have been common.
14 It is spread by the elm-bark beetle which carries the spores of the fungus with it. The fungus apparently originated in Asia and not Holland. A more virulent strain of the disease was introduced into Britain in the 1960s on logs imported from Canada.
15 Litter formed from dead materials such as corpses, dung, leaf litter etc.

Nature Quiz

1. What type of plant is Irish lady's tresses?
2. What is oology?
3. Most grasses are pollinated by wind. True or false?
4. What is an eagle's nest called?
5. What are fossil fuels?
6. When did life begin on earth?
7. Some areas of Antarctica were once covered with forests. True or false?
8. The word 'tundra' indicates what?
9. Do all conifers keep their leaves in winter?
10. Goldcrests build a deep cup-shaped nest made of mosses and lichens and bound with spiders' webs. How many eggs complete a clutch?
11. What is a gila monster?
12. What is the world's biggest frog?
13. A flea can jump 200 times its own height. True or false?
14. What is the difference between an Indian and an African rhinoceros?
15. What is the only female animal to have antlers?

ANSWERS

1. An orchid found in wet grassy places and marshy areas. It flowers in August and September.
2. The study of birds' eggs.
3. True.
4. An eyrie.
5. Deposits of organic matter, altered from pressure or decomposition, that are dug up and can be burnt as fuel. These include coal, natural gas and oil.
6. 3,600 million years ago.
7. True. Coal and fossil remains have been found.
8. High altitudes and high latitudes where vegetation is stunted and sparse.
9. No, some lose them. e.g. The larch.
10. Double-brooded, they normally have 7 to 10 eggs and occasionally as many as 13.
11. A lizard with a poisonous bite, found in North America.
12. The goliath frog of West Africa. It measures more than three-quarters of a metre (30 inches plus) and weighs around 3 kg. (7 lbs.).
13. True.
14. The Indian rhinoceros has only one horn.
15. The caribou.

Nature Quiz

1. What family of snake does the asp belong to?
2. What is pathology?
3. Bees have six wings. True or false?
4. What animal is traditionally known as 'the ship of the desert'?
5. What is a baby eagle called?
6. Which shaggy creature lives at very high altitudes in Tibet?
7. Where would you find most of the world's fresh water?
8. What natural substance is, weight for weight, stronger than steel, more elastic than rubber, and tougher than a bullet-proof waistcoat?
9. What creature, apart from man, puts 'glass' windows into its home?
10. What is plate tectonics?
11. What animals have built in life-jackets?
12. How energy-efficient is a firefly compared to a lightbulb?
13. What animal is seriously affecting the coral of the Great Barrier Reef?
14. How much water can a medium-sized oak tree draw up from the soil in a day?
15. What is the Richter scale?

ANSWERS

1. The cobra family.
2. The study of diseases.
3. False. Bees have four wings.
4. The camel.
5. Eaglets.
6. The yak.
7. In Antarctica, where 90% of the world's fresh water is locked up in ice.
8. A spider's web.
9. Certain species of tropical wasp use their saliva to form small hardened translucent panes. These they place in the outer shells of nests to make windows.
10. The study of the earth's crust based on the theory that the surface of our planet is made up of a number of continually moving and interacting plates.
11. Seals. Many seal species sleep upright in the water. To prevent themselves from sinking, they inflate their throats with air, which keeps them afloat in the same way as a lifejacket supports a human.
12. Most electric bulbs waste 97% of their energy in heat, whereas a firefly concentrates 90% of its effort into light.
13. The crown of thorns starfish.
14. As much as 140 gallons (637 litres) in a day. A tree's 'water pump' is powered by the sun. As water evaporates from the leaves, more is drawn up through tiny tubes in the trunk and branches. This is called evapotranspiration.
15. The scale used to measure the strength of earthquakes.

Nature Quiz

1. What plant's defence system is like a hypodermic syringe?
2. What percentage of the earth's surface is permanently covered in ice?
3. What is the longest mountain range in the world?
4. A dromedary has two humps. True or false?
5. How many animal species could face extinction as a result of the disappearance of a single plant. Is it: 5, 10, 20 or 30?
6. What does the ozone layer do?
7. Seaweeds come in three colours. What are they?
8. What are CFCs?
9. How far can a flying fish glide through the air?
10. Silkworms feed on the leaves of which tree?
11. What percentage of all species that have ever lived has already become extinct?
12. What are demersal fish?
13. All conifers bear cones. True or false?
14. A female cottontail rabbit can produce six litters of five young in a year. If all her offspring lived and reproduced, how big would her family be after six years?
15. All hard-woods are hard. True or false?

ANSWERS

1. The stinging nettle. The slightest contact causes the nettle hair to push into the skin at which point a bulb at the root of the hair squeezes a mixture of histamine and a poison similar to wasp venom into the victim.

2. Ten percent.

3. The Andes in South America.

4. False. The dromedary has only one hump whereas the camel has two.

5. Scientists have estimated that as many as 30 animals could become extinct.

6. It protects the earth from harmful short-wave radiation from outer space.

7. Red, brown and green.

8. Chlorofluorocarbons.

9. Up to 275 metres (300 yds.) at best. They glide at a height of 1.2 to 1.5 metres (4 to 5 feet) at speeds up to 65 km/h. (40 mph.).

10. The mulberry.

11. 90%.

12. Fish that feed on or near the bottom.

13. False. Some produce berry-like arils (like the yew berry).

14. 11 million cottontail rabbits! Thankfully the cottontail population is kept under control by disease and predators.

15. False. Balsa is a hard-wood.

Nature Quiz

1. The peacock, red admiral and small tortoiseshell butterflies all feed on the same plant. What is it?
2. What proportion of the world's plants, birds and animals live in tropical rainforests?
3. What plant covers the largest area of the world?
4. What is a 'thrush's anvil'?
5. What are young racoons called?
6. Where is the largest animal 'crèche' in the world?
7. What is the rhinoceros' horn made of?
8. Which is the biggest member of the cat family?
9. When two poplar aphids battle over possession of a poplar leaf, how long could the fight last?
10. Which bird is traditionally the symbol of peace?
11. How many tentacles does a squid have?
12. What is the difference between the black panther and the black leopard?
13. In olden days, people used quills to write with. Which bird provided them?
14. What bird is the national emblem of the United States?
15. What is a shepherd's purse?

ANSWERS

1. The stinging nettle.
2. 50%.
3. Plankton.
4. A hard object, often a rock or stone, used by a thrush to break open snail shells.
5. Kits.
6. The Carlsbad Cavern in New Mexico, USA, which is home to over 20 million Mexican free-tailed bats. Each night the young bats (about 2,000 to a square metre) are left hanging in the cave while their mothers fly off to feed.
7. Hair.
8. The tiger.
9. The resident aphid will try to kick and shove the intruder off and the intruder kicks and shoves back. The battle can go on for two or three days until one or the other prevails.
10. The dove.
11. Ten.
12. None, they are the same animal.
13. The grouse.
14. The bald eagle.
15. A common white-flowered plant of tilled ground and waste areas. It flowers throughout the year. The seed pods are green and triangular.

Nature Quiz

1. What is a little stint?
2. What is a sea potato?
3. Why do birds rarely nest in elder trees?
4. What is a bird's eggshell made of?
5. Why was the wild dog-rose so useful during World War 2?
6. A female perch can lay up to 200,000 eggs when spawning. True or false?
7. The shape of a swift's wings in flight is often described as what?
8. What colour is corncockle?
9. The wood white butterfly is now extinct in Ireland. True or false?
10. Most crabs 'laterigrade'. What does this describe?
11. What is a goldfinches' favourite food plant?
12. What is the difference between a Dublin Bay prawn and a Norway lobster?
13. What is zootoxin?
14. How did the liverwort get its name?
15. What is the largest freshwater crustacean in Ireland?

ANSWERS

1. A small wading bird seen regularly on autumn migration in esturaries around the country.
2. A sea urchin which differs from other sea urchins in that it lives in burrows on sandy shores near the low tide level.
3. The leaves have a particularly unpleasant smell, rather like that of a neglected mouse cage. It is thought that this is why birds avoid nesting in these trees.
4. Calcium carbonate and a small amount of organic matter.
5. The rose-hips are a valuable source of vitamin C. During the war, when citrus fruits were scarce, the hips were collected to make rose-hip syrup.
6. True, however this does depend on her size. Only the largest fish, weighing 2 kg. (4.5 lbs.) would lay as many as this.
7. Sickle-shaped. An apt description for this bird capable of speeds of over 96 km/h. (60 mph.) and perhaps even twice this in dives.
8. This plant's flowers are a rich purple colour. Once common in cornfields, it is now believed to be extinct in Ireland.
9. False. This attractive butterfly is still widely distributed throughout Ireland.
10. Literally 'walking sideways'.
11. Thistle. The goldfinch eats the seeds from mid-summer onwards.
12. None. They are the same species.
13. It is a toxin produced by an animal. e.g. Snake venom.
14. There are two types of liverwort. The name comes from the flat-lobed leaves of the thallose liverwort which vaguely resembles the human liver.
15. The freshwater crayfish. They grow to an average size of 10 cm. (4 inches).

Some Wildlife Rules & Laws

What are rules? Statements formulating what appear to be regularities in natural phenomena. As in most everyday situations there are exceptions to all rules and laws.

Allen's Rule (Proposed by J.A. Allen in 1876).
Warm-blooded species in a cold climate typically have shorter protruding body parts relative to body size than another race of the same species in a warm climate. This is because long protruding parts emit more body heat, and so are disadvantageous in a cool environment but advantageous in a warm environment.

Bergmann's Rule (Proposed by C. Bergmann in 1847).
Races of species from cold climates tend to be composed of individuals physically larger than those of races from warm climates. This is because the body surface area decreases as the body weight increases. A large body loses proportionately less heat than a small one. This is an advantage in a cold climate but disadvantageous in a warm one.

Buys Ballot's Law (Enunciated in 1857 by Christoph H. Buys Ballot).
In the northern hemisphere the winds blow anticlockwise around centres of low pressure and clockwise around centres of high pressure. In the southern hemisphere both these tendencies are reversed.

Gloger's Rule (Proposed by Constantin W.L. Gloger 1803-59).
Individuals of many species of mammals, birds and insects are darkly pigmented in humid climates and lightly coloured in dry ones. This may well be a camouflage adaptation.

Hesse's Rule (Proposed by R. Hesse 1921).
That among warm-blooded animals the forms living in cold regions have relatively higher heart weights than those living in warm regions.

Romer's Rule (Proposed by Alfred S. Romer 1894-1973).
The effect of many important evolutionary changes is to enable organisms to continue in the same way of life, rather than to adapt to a new one. i.e. The evolution of bony elements that strengthened the limbs of fish enabled them to crawl over land to find new ponds when the climate started to become dryer.

Wildlife Related Phobias

- Acarophobia — Fear of itching or of the insects that cause itching.
- Agrizoophobia — Fear of wild animals.
- Ailurophobia — Fear of cats.
- Alliumphobia — Fear of garlic.
- Ancraophobia — Fear of wind.
- Anthophobia — Fear of flowers.
- Anthrophobia — Fear of flowers.
- Antlophobia — Fear of floods.
- Apiphobia — Fear of bees.
- Arachnephobia — Fear of spiders.
- Arachnophobia — Fear of spiders.
- Astraphobia — Fear of lightning.
- Auroraphobia — Fear of aurora lights.
- Bacillophobia — Fear of microbes.
- Bacteriophobia — Fear of bacteria.
- Batonophobia — Fear of plants.
- Batrachophobia — Fear of amphibians, such as frogs, newts, salamanders, etc.
- Brontophobia — Fear of thunder.
- Bufonophobia — Fear of toads.
- Chinophobia — Fear of snow.
- Cremnophobia — Fear of precipices.
- Cryophobia — Fear of ice or frost.
- Cynophobia — Fear of dogs.
- Dendrophobia — Fear of trees.
- Doraphobia — Fear of fur or skins of animals.
- Entomophobia — Fear of insects.
- Gatophobia — Fear of cats.
- Heliophobia — Fear of the sun.
- Hellenologophobia — Fear of Greek terms or complex scientific terminology.
- Helminthophobia — Fear of worms.
- Herpetophobia — Fear of reptiles or creepy, crawly things.
- Hippophobia — Fear of horses.
- Homichlophobia — Fear of fog.
- Hydrophobia — Fear of water.
- Hygrophobia — Fear of dampness or moisture.
- Hylophobia — Fear of forests.

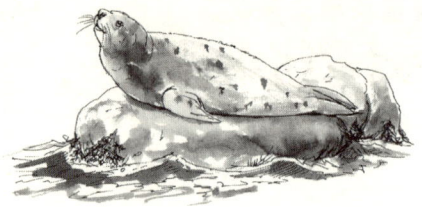

- Ichthyophobia — Fear of fish.
- Insectophobia — Fear of insects.
- Isopterophobia — Fear of termites, insects that eat wood.
- Keraunophobia — Fear of thunder.
- Limnophobia — Fear of lakes.
- Lutraphobia — Fear of otters.
- Melissophobia — Fear of bees.
- Meteorophobia — Fear of meteors.
- Mottephobia — Fear of moths.
- Murophobia — Fear of mice.
- Musophobia — Fear of mice.
- Mycophobia — Fear of, or aversion to, mushrooms.
- Myrmecophobia — Fear of ants.
- Nephophobia — Fear of clouds.
- Ombrophobia — Fear of rain.
- Ophidiophobia — Fear of snakes.
- Ophiophobia — Fear of snakes.
- Ornithophobia — Fear of birds.
- Ostraconophobia — Fear of shellfish.
- Pteronophobia — Fear of feathers.
- Parasitophobia — Fear of parasites.
- Pediculophobia — Fear of lice.
- Phthiriophobia — Fear of lice.
- Phyllophobia — Fear of leaves.
- Potamophobia — Fear of rivers.
- Ranidaphobia — Fear of frogs.
- Scoleciphobia — Fear of worms.
- Siderophobia — Fear of stars.
- Snakephobia — Fear of snakes.
- Spermatophobia — Fear of germs.
- Spermophobia — Fear of germs.
- Spheksophobia — Fear of wasps.
- Suriphobia — Fear of mice.
- Taeniophobia — Fear of tapeworms.
- Teniophobia — Fear of tapeworms.
- Thalassophobia — Fear of the sea.
- Verminophobia — Fear of germs.
- Zemmiphobia — Fear of the great mole rat.
- Zoophobia — Fear of animals.

Nouns of Multitude

BIRDS
Bar-tailed godwits	Ungodliness.
Birds	Congregation, Dissimulation (young), Flight, Flock, Volery, Volley.
Bitterns	Sedge, Siege.
Black-tailed godwits	Godliness.
Budgerigars	Chatter.
Bustard	Flock.
Capercaillie	Tok.
Choughs	Chattering.
Coots	Covert, Raft.
Cormorants	Flight.
Cranes	Herd, Siege.
Crows	Clan, Hover, Murder.
Curlews	Herd.
Dotteral	Trip.
Doves	Dole, Flight, Prettying.
Ducklings (in nest)	Clutch.
Ducklings (off nest)	Clatch.
Ducks (diving)	Dopping, Dropping.
Ducks (flying)	Flush, Plump, Skein, Team.
Ducks (on water)	Badeling, Paddling, Sail.
Eagles	Convocation.
Falcons	Cast.
Finches	Charm, Flight.
Flamingos	Flurry, Regiment, Skein.
Geese (flying)	Gaggle, Flock or Skein.
Geese (on land)	Gaggle.
Geese (on water)	Gaggle, Plump.
Goldfinches	Charm, Chattering, Chirp, Drum.
Goshawks	Flight.
Grouse	Brood, Covey, Pack.
Guillemots	Bazaar.
Gulls	Colony.
Hawks	Cast.
Herons	Scattering, Sedge, Siege, Rookery.
Ibis	Crowd.
Jays	Band, Party.
Lapwings	Deceit, Desert.
Larks	Exultation.
Magpies	Tiding, Tittering.
Mallard (on land)	Bord, Flock, Flush, Suite, Sute.
Mallard (on water)	Sord.
Nightingales	Match, Puddling, Watch.
Ostrich	Flock, Troop.
Owls	Parliament, Stare.
Oxbirds	Fling.

Parrots	Flock.
Partridge	Covey.
Passenger pigeons	Roost.
Peacocks	Muster.
Penguins	Colony, Rookery.
Pheasants	Brook, Ostentation, Pride, Nye.
Pigeons	Flight, Flock.
Plover	Congregation, Flight, Stand, Wing.
Ptarmigan	Covey.
Quail	Bevy, Covey.
Ravens	Unkindness.
Redwings	Crowd.
Rooks	Building, Clamour, Parliament, Rookery.
Ruffs	Hill.
Sandpipers	Fling.
Shelduck	Dapping, Dropping.
Skylarks	Exaltation.
Snipe	Walk, Whisper, Wish, Wisp.
Sparrows	Host, Surration, Quarrel.
Starlings	Chattering, Crowd, Murmuration.
Storks	Herd, Mustering.
Swallows	Flight.
Swans	Bank, Bevy, Drift, Game, Herd, Squadron, Teeme, Wedge, Whiteness.
Swifts	Flock.
Teal (on land)	Bunch, Coil, Knab, Raft.
Teal (rising)	Spring.
Thrush	Mutation.
Turtle doves	Pitying.
Wigeon	Coil, Company, Flight, Knob.
Wildfowl	Plump, Sord, Sute, Trip.
Woodcock	Covey, Fall, Flight, Plump.
Woodpeckers	Descent.
Wrens	Herd.

WILD ANIMALS

Animals	Menagerie, Tribe.
Antelope	Herd, Troop.
Apes	Shrewdness.
Baboons	Troop.
Badger	Cete, Colony.
Bears	Sleuthe, Sloth.
Beavers	Colony.
Bison	Herd.
Boars	Herd, Singular, Sounder.
Buffalo	Herd.
Camels	Caravan, Flock.
Chamois	Herd.
Deer	Herd, Leash.
Dolphins	Pod, School
Elephants	Herd
Elephant seals	Rookery, Team, Troop.
Elk	Gang.
Foxes	Earth, Lead, Skulke.
Giraffes	Corps, Herd, Troop.
Hares	Down, Drove, Husk, Lie, Trip.
Hart	Herd, Stud.
Hedgehogs	Array.
Hippopotamuses	Herd, School.

Hogs	Herd, Drove, Sounder.
Kangaroos	Herd, Mob, Troop.
Lemurs	Troop.
Leopards	Leap.
Lions	Flock, Pride, Sawt, Souse, Troop.
Martens	Raches, Richesse, Richness.
Mice	Nest.
Moles	Company, Labour, Movement, Mumble.
Monkeys	Tribe, Troop.
Moose	Gang, Herd.
Otters	Bevy, Family.
Oxen (wild)	Drove, Herd.
Pine martens	Raches, Richesse, Richness.
Polecats	Chine.
Porpoises	Gam, Pod, School.
Rabbits	Bury, Colony, Nest, Warren.
Racoons	Nursery.
Rats	Colony.
Rhinoceroses	Crash.
Roe Deer	Bevy.
Seals	Harem, Herd, Pod, Rookery.
Squirrels	Drey.
Stoats	Pack.
Swine (wild)	Sounder.
Tigers	Ambush.
Walrus	Herd, Pod.
Weasels	Pack, Pop.
Whales	Colony, Gam, Herd, Pod, School.
Wild cats	Dout.
Wild cattle	Mob.
Wolves	Pack, Rout.
Zebras	Herd, Kinship.

DOMESTIC ANIMALS & BIRDS

Asses	Herd, Pace.
Bloodhounds	Sute.
Cats	Chowder, Clowder, Cluster.
Cattle	Drove, Herd.
Chickens	Brood, Clutch, Peep.
Colts	Race, Rag, Rake.
Dogs	Cowardice, Kennel, Pack.
Dogs (hunting)	Cry.
Donkeys	Herd, Drove.
Ferrets	Business, Cast, Fesynes.
Foxhounds	Pack.
Goats	Flock, Herd, Tribe, Trippe.
Greyhounds	Brace, Leash, Pack.
Hens	Brood, Flock.
Horses	Harass, Herd, Stable, Stud, Troop.
Horses (race)	Stable, String.
Hounds	Brace, Couple, Cry, Mute, Pack, Stable.
Kittens	Brood, Kindle, Litter.
Mares	Flock, Stud.
Mules	Barren, Cartload, Pack, Span.
Oxen	Drove, Rake, Team, Yoke.
Peafowl	Muster, Ostentation, Pride.
Piglets	Farrow.
Pigs	Litter, Herd, Sounder.
Ponies	Herd.

Poultry	Flock, Run.
Pups	Litter.
Racehorses	Field, String.
Sheep	Down, Drove, Flock, Hurtle, Trip.
Spaniels	Couple.
Swine	Doyet, Dryft.
Turkeys	Dule, Raffle, Rafter.

PLANTS

Nuts	Cluster.
Trees	Clump, Coppice, Wood.

AMPHIBIANS & REPTILES

Frogs	Army, Colony.
Snakes	Den, Pit.
Snakes (young)	Bed.
Toads	Knab, Knot.
Turtles	Bale, Dole.
Vipers	Den, Nest.

INVERTEBRATES

Ants	Army, Column, State, Swarm.
Bees	Cluster, Erst, Hive, Swarm.
Caterpillars	Army.
Flies	Business, Cloud, Scraw, Swarm.
Gnats	Cloud, Horde, Swarm.
Grasshoppers	Cloud.
Insects	Swarm.
Lice	Flock.
Locusts	Cloud, Horde, Plague, Swarm.
Spiders	Cluster, Clutter.
Wasps	Herd, Nest, Pladge.

FISH

Angel fish	Host.
Barracuda	Battery.
Bass	Fleet.
Clams	Bed.
Cockles	Bed.
Dog fish	Brood, Troop.
Eels	Swarm.
Fish	Haul, Run, School, Shoal.
Goldfish	Troubling.
Herring	Army, Gleam, Shoal.
Jellyfish	Brood, Smuck.
Mackerel	School, Shoal.
Minnows	Shoal, Steam, Swarm.
Mussels	Bed.
Oyster	Bed.
Perch	Pack, Shoal.
Pilchards	Shoal.
Roach	Bevy.
Salmon	Bind.
Sardines	Family.
Smelt	Quantity.
Sticklebacks	Shoal.
Trout	Hover.
Whiting	Pod.

Some Useful Wildlife Contact Addresses

Artists for Nature Foundation
Address: Striepeweg 5, 7675 TG Bruinehaar, The Netherlands.
Telephone: 0031-546-681-253.

BirdWatch Ireland
Address: Ruttledge House, 8 Longford Place, Monkstown, Co Dublin.
Telephone: 01-2804322.

British Trust for Ornithology
Address: The Nunnery, Nunnery Place, Thetford, Norfolk, IP24 2PU.
Telephone: 0044-1842-750050.

Castle Espie, Wildfowl and Wetlands Trust Reserve
Address: Ballydrain Road, Comber, Co. Down, BT23 6EA.
Telephone: 08-01247-874146.

Conservation Volunteers Ireland
Address: P.O.Box 3836, Ballsbridge, Dublin 4.
Telephone: 01-6681844.

Earthwatch
Address: Harbour View, Bantry, County Cork.
Telephone: 027-50968.

ECO
Address: 26 Clare Street, Dublin 2.
Telephone: 01-6625491.

ENFO
Address: 17 Andrew Street, Dublin 2.
Telephone: 01-6793144.

Field Studies Council
Address: Central Services, Preston Montford, Shrewsbury,
            Shropshire SY4 1HW.
Telephone: 0044-1743-850674.

Irish Peatland Conservation Council
Address: 119 Capel Street, Dublin 2.

Irish Naturalists' Journal
Address: Ulster Museum, Botanic Gardens, Belfast, BT9 5AB.

Irish Raptor Study Group
Address: c/o Ruttledge House, 8 Longford Place, Monkstown, Co Dublin.

Irish Seal Sanctuary
Address: Garristown, Co. Dublin.
Telephone: 01-8354370.

Irish Wildlife Trust
Address: 107 Lower Baggot Street, Dublin 2.
Telephone: 01-6768588.

Marine Strandings Project
Address: Irish Whale and Dolphin Group, University College, Cork.
Telephone: 021-904197 / 904053.

Met Éireann
Address: Glasnevin Hill, Dublin 9.
Telephone: 01-8424411.

National Parks and Wildlife, Dúchas The Heritage Service
Address: Mespil Road, Dublin 4.
Telephone: 01-6670788.

Royal Society for the Protection of Birds
Address: The Lodge, Sandy, Bedfordshire, SG19 2DL.
Telephone: 0044-1767-680551.

Tree Council of Ireland
Address: The Royal Hospital, Kilmainham, Dublin 8.
Telephone: 01-6970699.

Wexford Wildfowl Reserve
Address: North Slob, Wexford.
Telephone: 053-23129.